CARDIFF CLOCKS

CARDIFF CLOCKS

Clock and Watchmakers of Cardiff, the Vale of Glamorgan and the Valleys

BY

WILLIAM LINNARD

MERTON PRIORY PRESS

First published 1999

Published by
Merton Priory Press Ltd
67 Merthyr Road, Whitchurch
Cardiff CF4 1DD

© William Linnard 1999

ISBN 1 898937 29 X

Printed by
Hillman Printers (Frome) Ltd
Handlemaker Road, Marston Trading Estate
Frome, Somerset BA11 4RW

Contents

Introduction	9
Clocks in South Wales: a Brief Historical Survey	11
Clockmakers of Cardiff, the Vale and the Valleys	22
Features of Local Longcase Clocks	39
Cases	39
Movements	43
Dials	46
Signatures	50
Local Clocks on Public Display	55
Museum of Welsh Life, St Fagans	58
Cyfarthfa Castle Museum and Art Gallery, Merthyr Tydfil	58
Pontypridd Historical and Cultural Centre	60
Tredegar House, Newport	60
Newport Museum and Art Gallery	61
Torfaen Museum Trust, Pontypool	63
Select Bibliography	64
Comprehensive List of the Clockmakers of Cardiff, the Vale and the Valleys	65
Appendix. Facsimile of the Catalogue and Price List issued by Howell Morgan of Blaengarw, *c.* 1900	81

CLOCK-*maker*. If we were to define the word clock-maker agreeably to the derivation of the term, we should simply say that it means a man who makes clocks, and this definition, at one period of the art, would have been sufficient for our purpose; but since clocks have become so common as to be considered as articles of household furniture, the art of making them has not been confined, as at first, to one department of mechanics, but has gradually ramified into various branches, so distinct from one another, that the maker of one part is frequently unacquainted with the operations requisite for the manipulations of another, equally essential.

Abraham Rees, *New Cyclopaedia* (1819)

Fig. 1. South Wales, showing places mentioned in the text (*Pigot's Directory of North and South Wales*, 1844).

INTRODUCTION

DR IORWERTH PEATE'S BOOK *Clock and Watch Makers in Wales* (1945), and the later revised and expanded editions of 1960 and 1975, have formed the foundation for all subsequent studies of Welsh horology. Since then, and building on Peate's pioneering work, other writers have made significant contributions to horology in various parts of Wales. Apart from numerous published articles, several important books have also appeared: a detailed study of an individual clockmaker in mid Wales (W.T.R. Pryce and T. Alun Davies, *Samuel Roberts Clock Maker*, 1985), those of a single North Wales town (Colin Brown and Mary Brown, *The Clockmakers of Llanrwst,* 1993), and those of a major area within North Wales (Paul Parker, *Clockmaking in the Vale of Clwyd*, 1993).

In South Wales, Peter Hurst Jarvis's unpublished dissertation (University of Wales, Cardiff 1995) entitled 'Clockmaking in Glamorgan in the Eighteenth and Nineteenth Centuries' is a most useful local history. More recently still, a major contribution on part of South Wales has appeared, namely Joanna Greenlaw's *Swansea Clocks: Watch and Clockmakers of Swansea and District* (1997). This book covers Swansea and its hinterland, an arc embracing Llanelli, Llandeilo and Neath.

All this has provided a fuller picture of horological activity in much of Wales during the eighteenth and nineteenth centuries. However, one obvious gap in the coverage still exists, namely Cardiff, the capital city of Wales, together with its rural and industrial hinterland. Accordingly, this book seeks to describe the clock- and watch-makers of Cardiff, the Vale of Glamorgan and the industrial valleys up to 1900. It covers the area from Bridgend to Newport, and includes the towns of Aberdare, Merthyr Tydfil, Pontypridd and Pontypool.

It is a most pleasant duty to record my thanks to a number of individuals and institutions, who have assisted me in the compilation of this book: Joanna Greenlaw (author of *Swansea Clocks*), who pointed the way; Chris Roberts (clock restorer and repairer, Cardiff), who provided much sound, practical advice and information; Dr Fred Holley, the Merthyr historian; and historical geographer Dr W.T.R.

Pryce of the Open University in Wales, Cardiff, who very kindly read a draft of this book and made many valuable comments and improvements.

Staff of the following institutions have kindly facilitated access to the records or the clocks (or both) in their care: Museum of Welsh Life, St Fagans; Cardiff Central Library; Cardiff Castle; Glamorgan Record Office, Cardiff; Swansea Record Office; Pontypridd Historical and Cultural Centre; Merthyr Tydfil Library; Cyfarthfa Castle Museum and Art Gallery, Merthyr Tydfil; Tredegar House, Newport; Gwent Record Office (Cwmbran); Torfaen Museum Trust Ltd (Pontypool); Radyr Library; and Newport Museum & Art Gallery.

Finally, I owe a great debt of gratitude to Mr Glyn Jenkins for preparing the manuscript for publication.

Of course, I am solely responsible for any errors of omission or commission. I am well aware that a work of this kind can never be the 'last word' on the subject, and would welcome constructive comments and new information, which may be sent via the publisher.

March 1999   William Linnard

CLOCKS IN SOUTH WALES
A BRIEF HISTORICAL SURVEY

> Och i'r cloc yn ochr y clawdd
> Du ei ffriw, a'm deffroawdd
> Difwyn fo'i ben a'i dafod
> A'i ddwy raff iddo, a'i rod,
> A'i bwysau, pelennau pŵl
>   Dafydd ap Gwilym, *fl.* 1340–70, *Cywydd y Cloc*

THE SHORTAGE OF DOCUMENTARY SOURCES, coupled with the lack of really early surviving timekeepers, make the earliest history of clocks in Wales a matter of conjecture.

The Account Book of Beaulieu Abbey in Hampshire records that a 'horologiarius' was employed for five whole weeks at the sacristy of the Abbey during the financial year Michaelmas 1269 to Michaelmas 1270, evidently a specialist working on some form of timekeeper (*horologium*).[1] It is tempting, but pointless, to speculate whether this *horologium* was in fact a *clepsydra* (water clock), a sundial, or even perhaps an extremely early mechanical clock. Whatever form of timekeeper it was, it is more than likely that the great Cistercian houses of South Wales at Tintern, Margam and Neath would have been well aware of horological developments at their sister abbey at Beaulieu, and might even have sought to emulate them. Unfortunately, nothing survives that might shed any light on horology in this early period in Wales.

The earliest description of a clock in Wales is the famous, but tantalising, *cywydd* by Dafydd ap Gwilym (*fl.* 1340–70) quoted in part at the head of this page. This poem has been frequently quoted, variously translated, and differently interpreted. The great turret clock which Dafydd ap Gwilym described in such detail in the fourteenth century may have been located at Brecon, Llanthony, somewhere else in South Wales, somewhere in England, or even at Rouen, as has

---

[1] S.F. Hockey (ed.), *The account book of Beaulieu Abbey* (Camden 4th Series, xvi, Royal Historical Society, 1975), pp. 34, 235, 240, 299.

been argued by Gareth Evans.[1] Wherever the famous black-faced clock was, Dafydd ap Gwilym's poem certainly bears all the marks of being a first-hand, eye-witness description of a turret clock, with its ropes, wheel, weights, heavy balls, hammer, and so on.

If mechanical clocks were great rarities in the time of Dafydd ap Gwilym, after the fourteenth century they gradually became more common, although initially only for major ecclesiastical buildings such as cathedrals and churches, public buildings such as town halls, or as the prized possessions of a few wealthy individuals. St David's Cathedral certainly had a clock in the fifteenth century if not earlier, and Iorwerth Peate summarised several poetical references to time-keepers (sun-dials, watches and clocks) in Wales in the fifteenth and sixteenth centuries.[2] There was a 'clockhouse' at Caerleon in 1618.[3]

One of the earliest Welsh clocks for which quite good documentary accounts survive is the town clock in Haverfordwest. The published Haverfordwest town accounts for 1563–1600 indicate that this clock was certainly installed and functioning before the beginning of this period. Payments are recorded for wires, oil, boards and 'other things for the clock' at various dates between 1564 and 1600, and cash was paid to several persons for minor and major repairs to the clock during that period: John Webb, Rice Tynker, an unnamed cooper, Lewis Webb and Richard Harris (smith). This town clock was very important in the commercial life of Haverfordwest, where local regulations stated unambiguously that 'no flesh or fish or other victual be bought before viii at the clock'; that 'no corn be bought before the hour of xi of the clock'; and that 'no foreigner shall buy corn before xii of the clock'.[4]

The comprehensive inventory of the goods of Sir John Perrot in Carew Castle (Pembs.) in 1592 listed among the 'brasse, laten, copper, ledd and ironstuff' the following item: 'a clock, price xs'.[5]

Surviving parish records from places as diverse as Neath in Glamorgan and Meifod in Montgomeryshire give glimpses of the installation and maintenance of turret clocks in Welsh churches. At

---

[1] *Y Traethodydd*, cxxxviii (1982), 7–16.

[2] I.C. Peate, *Clock and watch makers in Wales* (1975), pp. 16ff.

[3] Gwent Record Office, D 580.170.

[4] B.G. Charles (ed.), *Calendar of the records of Haverfordwest 1539–1600* (University of Wales Press, 1967), pp. 24–5.

[5] *Archaeologia Cambrensis* (1866), p. 346.

St Thomas's Church, Neath, a clock was installed in 1695–6.[1] Meifod parish church even had a clock in the first half of the sixteenth century, and later churchwardens' accounts give details of expenditure and work on the Meifod clock from 1734 onwards.[2] Similar activity must have been going on at other churches in Wales, but was either unrecorded or the relevant records have not survived.

The great turret clock at the stables of Tredegar House, Newport, was reputed to have struck the quarters by boys (i.e. mechanical jacks), like the old clock formerly at St Dunstan's near Temple Bar in London. It is not known precisely when the Tredegar stable clock was first installed; it was replaced in 1766.[3]

A prominent Welsh horological experimenter in the seventeenth century was John Jones of Pentyrch, near Cardiff. Jones became Chancellor of Llandaf Cathedral in 1691 and is buried there. As a young man at Oxford, Jones had invented an ingenious but ultimately unsuccessful clock which came to the notice of Robert Plot, who described it in some detail:

> Amongst other *Aerotechnicks,* here is a Clock lately contrived by the ingenious *John Jones* LL.B. and Fellow of *Jesus College Oxon*: which moves by the *air*, equally expressed out of *bellows* of a *cylindrical* form, falling into folds in its descent, much after the manner of Paper *Lanterns:* These, in place of drawing up the weights of other *Clocks,* are only filled with *air*, admitted into them at a large orifice at the top, which is stop'd up again as soon as they are full with a hollow *screw*, in the head whereof there is set a small *brassplate*, about the bigness of a silver halfpenny, with a hole perforated scarce so big as the smallest pins head: through this little hole the *air* is equally expressed by *weights* laid on top of the *bellows*, which descending very slowly, draw a *Clock-line,* having a counterpoise at the other end, that turns a pully-wheel, fastened to the *arbor* or *axis* of the *hand* that points to the *hour:* which device, though not brought to the intended perfection of the Inventor, that perhaps it may be by the help of a *tumbrel* or fusie, yet highly deserves

---

[1] *Trans. Neath Antiquarian Society* (1978), pp. 35–47.

[2] W.T.R. Pryce and T.A. Davies, *Samuel Roberts, Clock Maker: an eighteenth-century craftsman in a Welsh rural community* (1985), pp. 31–7.

[3] *Arch. Camb.* (1886), p. 102.

mentioning, there being nothing of this nature that I can find amongst the writers of *Mechanicks*.[1]

The Worshipful Company of Clockmakers of the City of London was formed in 1631 and thereafter numbers of Welshmen went to London to serve their apprenticeships.[2] Some stayed on in the city, other returned home to set up business; other men of a practical and enquiring nature taught themselves the principles of clockwork. Towards the end of the seventeenth century several individual clockmakers became established in South Wales. It must be remembered, however, that the population of Wales was too low to support many clockmakers. The population of Glamorgan in 1750 is estimated at only 55,200, while that of Monmouthshire was perhaps 40,600.[3]

The earliest clockmaker known in Chepstow was a man called Taynton who worked there in the 1660s. In addition to smoke-driven spits he made 'a clock that went with Air, and another with Water. And ... a very small Watch, most part of it being only of Wood'. Taynton was assisted for a time by his stepson William Bedloe.[4] According to Peate, a certain Humphrey Morgan made clocks in Abergavenny in the late seventeenth century, but in their analysis of his list, Pryce and Davies incorrectly placed this man in Llandaf.[5] Nothing further is known of him.

Individual clockmakers definitely known to have been active in the Vale of Glamorgan before the end of the seventeenth century included Griffith Jenkins of Llantrisant and William Evan Water (i.e. Walter) of St Mellons.[6] In the early years of the eighteenth century many more can be identified.

The oldest domestic clocks made in Glamorgan, either actually surviving or at least reliably documented, appear to be those made by Samuel Verrier of Pontneddfechan and Henry Williams of Llancarfan —significantly both men who had come from England to settle in Wales—and others such as Thomas French of Wenvoe, Philip Walton

---

[1] R. Plot, *Natural History of Oxfordshire* (1677), p. 230.

[2] Peate, *Clock and watch makers*, p. 88.

[3] J. Williams, *Digest of Welsh historical statistics* (Welsh Office, 1985), p. 6.

[4] I. Waters, *Chepstow Clock and Watch Makers* (1978), pp. 5–6.

[5] Pryce and Davies, *Samuel Roberts*, p. 9.

[6] See the List of Makers for further details.

of Cowbridge, and Edmond Edmonds of Laleston. In the mid eighteenth century Charles Vaughan of Pontypool and Griffith Williams of Newport also produced many brass-dial longcase clocks. Pryce and Davies mapped and described the development of clockmaking in Wales: only eight clockmakers were known in the whole of the country before 1700. By 1750 the figure had risen to 72, and by 1800 to well over two hundred. Thereafter, numbers increased rapidly during the nineteenth century, with concentrations in all the major towns.[1] Recent research has identified a few more clockmakers active at the end of the seventeenth century, still more in the eighteenth century, and many more in the nineteenth, and their names are included in the list at the end of this book.

Ownership of domestic clocks and watches by well-to-do private individuals in Wales gradually became more common during the seventeenth century. For example, by the 1730s there were four clocks in the mansion of Cefn Mabli (Glam.).[2] Although Bob Owen suggested that clocks and watches rarely featured in Welsh wills and inventories before about 1740,[3] they are mentioned in some earlier Sully inventories, including those of Moore Perkins in 1681 ('One Clock & its Appurtenances' £2), Christopher Portrey in 1711 ('an old clock'), William Jenkins in 1728 ('a clock' 10s.). and Francis Bowen in 1735 ('one clock' £2 and 'two silver watches' 10s.).[4] In 1707 Jane Herbert of White Friars, Cardiff, left her gold watch to her goddaughter Judith Powell (daughter of Sir Thomas Powell); in 1716 Christopher Matthews, alderman of Cardiff, left a 'clock with its case' to his son William; and in 1774 Catherine Evans of St John's, Cardiff, left her silver watch to her son Edward.[5] There is no indication whether any of these timekeepers were actually made in South Wales.

The eighteenth century saw the great development of individual craftsmanship in clockmaking throughout Wales, and it is generally agreed that up until the end of that century a clockmaker would have designed and made his clocks by assembling together parts most of

---

[1] Pryce and Davies, *Samuel Roberts*, pp. 10–13.

[2] Gwent Record Office, MS Newport 6018.

[3] B. Owen, *Diwydiannau Coll* ('Vanished Industries') (1943), p. 79.

[4] G.M. Jones and E. Scourfield, *Sully* (1986), pp. 42, 45, 73.

[5] Extracted from inventories in J.H. Matthews (ed.), *Cardiff Records*, III, pp. 145, 157, 183.

which he himself had made. Thereafter it became possible, and increasingly advantageous in terms of costs, to buy semi-finished components and eventually complete movements from specialist workshops in Birmingham, London or south Lancashire, and the economic conditions of the trade became such that clockmakers, even in the most rural areas, increasingly chose to buy in parts in preference to manufacturing them.

In 1797 a tax was introduced on clocks and watches. This was bitterly resented and was repealed in 1798, but it did increase demand for cheap clocks for inns and other public places. Such wall clocks became known as Act of Parliament clocks, a typical one by John Thackwell (Cardiff), *c.* 1800, having a plain white dial 16 inches in diameter.[1]

As early as 1819 Abraham Rees, himself a Welshman, a native of Llanbryn-mair, was able to state in his authoritative *New Cyclopaedia* that:

> a finisher of a clock now has no occasion to cast or cut his wheels himself, much less to make his springs or enamel his dial-plate. From custom, however, that man is called a clockmaker, who finishes or puts together the different constituent parts of a clock when made, and who has his profit from the sale of the machine; though the makers, more properly speaking, are the workmen employed in making the frame and contained wheel-work. The different operations may, indeed, be most of them performed by one workman, when the construction is intended to be peculiar, or the works of superior accuracy, but in general the different departments of the art may be separately enumerated, agreeably to the subjoined order, *viz.*
>
> The brass-founder crafts the wheels, plates, pillars, and faces, according to approved models:
> The spring-maker forges, shapes, and tempers the main-springs, to any required strength or dimensions:
> The making of the weights, to be used as maintaining powers of the balls, or bobs, and hands, may be considered as one branch:

---

[1] Peate, *Watch and clock makers*, p. 100.

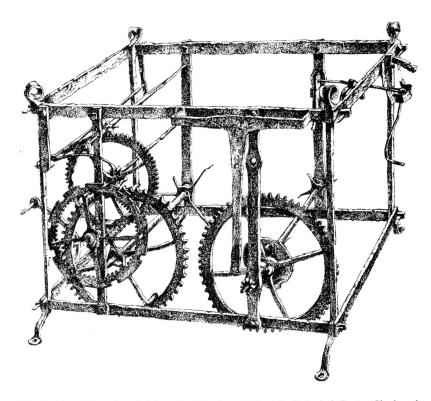

Fig. 2. *Above:* Remains of old turret clock from St David's Cathedral (Peate, *Clock and watchmakers*). *Below:* Cheap imported clocks are illustrated in popular printed matter from the mid nineteenth century, as on the title page of a religious poem, 'Meditation on the Clock Striking' *(Cardiff Central Library, Ballad Collection)*.

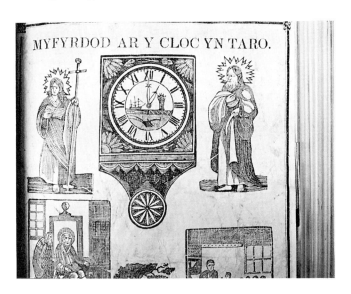

The man who keeps a cutting-engine and a fusee-engine, cuts the wheels and pinions, and forms the grooves on the fusee or barrel, accordingly as a spring or suspended weight is used as a maintaining power:

The movement-maker mounts the frame, makes the wheels, pinions, detents, &c. and places them in the frame, agreeably to the proposed calliper:

The clock-smith forges the steel pieces for the arbors, pinions, pallets, rack, hammer, detents, &c.:

The bell-founder casts the bell, or bells when the clock has chimes:

The enameller prepares the ground of the dial, or face, for receiving the colour of the figures, and gets the painter to lay on the figures, agreeably to the calliper, with or without a circle for the seconds:

When the face is not of real enamel, a japanner, or imitative enameller, prepares and finishes the dial:

When the face is brass silvered, an engraver usually prepares, and sometimes also silvers it:

A jeweller is employed for the pallets and pivot-holes of the best astronomical clocks and regulators:

The gilder is frequently employed for preparing the ornamental parts of the case:

The glazier is applied to for the door of the superior part of the case, when a seconds pendulum is used, and for the principal door sometimes, when the clock has a short pendulum:

The cabinet maker is resorted to, usually, for the case of the clock; and sometimes also the carver:

The chain or cat-gut maker is indispensably necessary:

Recently the tubular compensation-pendulum has been made and adjusted, by the mathematical instrument-maker, as being a portion that requires great precision:

Lastly, the finisher, or, as he is otherwise called, the *maker*, polishes the teeth and steel parts, finishes the pivots, verifies the engagement, adjusts the escapement, limits the arc of vibration by adjusting the maintaining power to the weight of the ball, regulates the adjustments for beat and rate, finishes the striking and repeating parts, and puts the whole machine into a state for sale.

HISTORICAL SURVEY 19

Fig. 3. *Above:* The Town Hall in High Street, Cardiff, in the 1850s *(Cardiff Central Library). Below:* The turret clock in the Town Hall maintained by William Wilson in the 1790s (Peate, *Clock and watchmakers*).

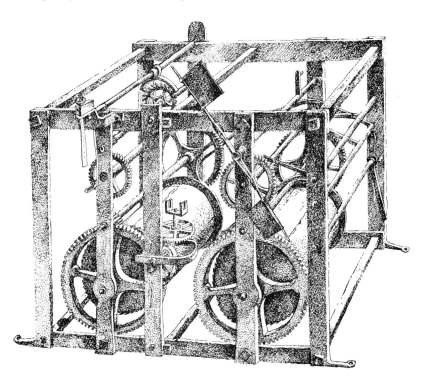

As Pryce and Davies summarised the situation, by about 1825 'persons described as clock and watchmakers in trade directories and official records ... were increasingly *former* clockmakers, or their descendants, who now concentrated on the retail side, undertaking repairs as well as selling factory-made clocks and watches'.[1]

This change in the trade from individual craftsmen to mass production was graphically described by that most accurate of observers, Alfred Russel Wallace, Darwin's rival in the theory of evolution. In his autobiography Wallace, who was born at Usk and lived and worked for some years in South Wales, and who had also worked briefly for a watchmaker in a small country town in England, recalled his escape from the monotony of the wholesale business in the 1840s:

> The movements of watches dealt in were purchased in Coventry, where the various kinds in general use were designed, the separate parts cast, machine-cut, and filed to their proper gauges, and put together. The mainsprings and balance-springs, chains, hands, dials and cases were usually purchased separately; and for each class of watch a fitter was employed, whose business it was to put the parts together, find out any small defects, and correct them by hand, while any larger defect in any particular part was sent back to the workman or manufacturer responsible for it. The man at the office made a final examination of the completed watches, tested their performance, corrected any minute defect that was discoverable, and finally, in consultation with one of the firm, determined the grade or quality of the watch and the consequent price. What I should have learnt there would have been how to fit a watch together, how to test it for definite defects, how to judge of the design and workmanship, how to keep accounts, pay the workmen, and probably to act as a traveller for the firm.[2]

Despite the proliferation of mass-produced clocks and watches during the nineteenth century, it is salutary to remember that many country folk were still experiencing difficulty in telling the time

---

[1] Pryce and Davies, *Samuel Roberts*, p. 13.
[2] A.R. Wallace, *My Life* (1905), p. 138.

during the hours of darkness. Ready access to light is now taken for granted, but even in the middle of the last century, people still depended on special measures to tell the time at night. Repeater devices were fitted on clocks to enable the owner to tell the last hour struck during the hours of darkness, and in 1840 William Hughes advertised the following home-made device to help Welshmen to tell the time by their watch in the dark:

> Take a small bottle of clear glass and place in it a spot of phosphorus the size of a pea, then fill it half full with boiled oil, cork it and leave it for a day or two. When you want to use it, hold it in your hands for a little time to warm up, then remove the cork for a moment, replace it quickly, and it will give enough light to be able to make out the figures and the hands.[1]

During the nineteenth century, rapid industrialisation, mass in-migration and population growth completely transformed the landscape, economy, society and language of South Wales. Against the background of these developments, the individual craft of making clocks and watches declined locally, and throughout Great Britain, as former makers resorted to repairing and also to the retailing of imported mass-produced clocks and watches manufactured in England, France, Germany and the USA. These trends are described in detail in the following chapters for Cardiff and its hinterland, the Vale of Glamorgan, and the industrial valleys of Glamorgan and Monmouthshire. This area extends from the river Garw in the west to the Usk in the east, and includes the towns and villages in the valleys of the rivers Ogwr, Ely, Taff, the two Rhonddas, Cynon, Rhymney, Sirhowy, Ebbw and Llwyd (see map).

The wide range of imported clocks and watches available in South Wales by the end of the nineteenth century is shown in the catalogue, reproduced in the Appendix.

---

[1] Translation from William Hughes, *Cyfaill y Cywrain* (1840), p. 45.

CLOCKMAKERS OF CARDIFF, THE VALE AND THE VALLEYS

> a very ingenious man ... and the first man
> of Knowledge in our parts in Clock work
> William Thomas, *Diary,* 1766

AFTER JOHN JONES OF PENTYRCH, who had produced his experimental bellows clock at Oxford before 1676, the earliest identifiable local clockmakers are rather shadowy figures about whom little is known. Of these, the first appears to be Griffith Jenkins, a clockmaker of Llantrisant who was admitted as a freeman of the borough. He died at Llantrisant in December 1729, leaving a total estate valued at £12 7s. 6d.; his tools included a pair of bellows, an anvil, a 'backorn', hammer and tongs, two vices, a small vice, and the rest of the implements belonging to his trade (valued at £3 17s. 2d. in all).[1] William Harrys (Harris), a clockmaker, was a prisoner in the county goal at Cardiff on 22 July 1734, the day his own son was baptised.[2] Beyond this fact, nothing is known of Harris or his clocks.

Other clockmakers working in the Cardiff district are identified in the pages of the extensive diaries kept by William Thomas of Michaelston-super-Ely from 1750 to 1795. The earliest of these appears to be William Evan Water (Walter) of St Mellons, born *c.* 1679.[3] This man was probably related to the Walter William who in 1736 presented his annual bill for clock maintenance and other work to Lady Jane Tynte of Cefn Mabli for the year 1 May 1735 to 1 May 1736:[4]

---

[1] J.B. Davies, *The freemen and ancient borough of Llantrisant* (1989), p. 67.

[2] *Cardiff Records*, III, p. 428.

[3] R.T.W. Denning (ed.), *The Diary of William Thomas 1762–1795* (South Wales Record Society, 1995), p. 171.

[4] Gwent Record Office, MS Newport 6018.

for keeping the Hall's Clock in order	0	5	0
for keeping the Gallerie's Clock	0	5	0
for keeping the Clock at the Servant's Hall	0	2	6
for keeping the Gate House Clock	0	5	0

William Thomas described him as 'a very ingenious man once and the first man of Knowledge in our parts in Clock work', but he died a pauper in 1766 'a drunken Ruinated man ... blind this ten years or more'. His son Harry, also a clockmaker of St Mellons, had 'fled to Portugal' years before.[1]

Thomas French (1701–93) of Wenvoe was 'a very knowing man in several branches of Knowledge, as clock and watch work, etc. Also a fiddler and by Trade a Glazier'. His son John French, also a clockmaker at Wenvoe, died in 1780, aged 52.[2]

Samuel Varrier (or Verrier) (c. 1695–1765) was a clockmaker and 'an Englishman by birth' who had worked in the Vale of Glamorgan in the area of St Fagans, Pentrebane and Llanfair (St Mary Church) for about twenty years, but later the old man used to come to the Vale every year from his son's place at Pontneddfechan on a regular circuit to see 'his friends, and others that he looked to their clocks by the year, the time of his fees near'. William Thomas described him as a 'short, currish sort of man ... Honest in his dealings'. Samuel Verrier actually died on one of these annual clock-maintenance tours, and was buried at St Fagans on 9 November 1765, aged 70.[3]

Others mentioned by name in William Thomas's Diary are Henry Williams of Llancarfan 'a Clock and a Watchmaker and a great farmer ... by birth from Gloucestershire' (died July 1790, aged 63); John Beynon and Jeffrey Beynon, both watchmakers, who were active in the Llandaf and Barry areas in the 1770s and 1780s; and a man called Phillip, 'a clockmaker at Pentyrch ... not from our Shire but ... very Ingenious'. This was David Phillip, or his brother John, both clockmakers.[4]

---

[1] *William Thomas's Diary*, p. 171.

[2] Ibid., pp. 129, 294.

[3] Ibid., pp. 149–50.

[4] Ibid., pp. 252, 331, 390.

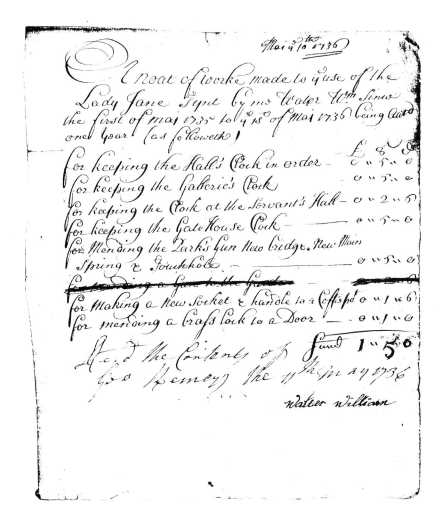

Fig. 4. Bill presented by Walter William for clock maintenance at Cefn Mabli in 1736 *(Gwent Record Office, MS Newport 6018)*.

William Thomas also refers to a rather nefarious character, 'William the Gingerbread Maker', born in Llancarfan, the son of a Bristol clockmaker, describing him as 'a man very deceitful with his creditors—taking the feathers from the beds and the clocks out of the cases'.[1]

The churchwardens' accounts for St John's Church, Cardiff, refer to the church clock from 1739 onwards. Several local clockmakers are mentioned in connection with work on this clock: Henry Williams (Llancarfan) in 1776 and 1780, John Thackwell in 1785, William Wilson in 1790, and Thackwell again in 1814 and subsequently.[2]

Another local Vale diarist, John Perkins of Llantrithyd, referred several times during the 1780s and 1790s to Edward (Ned) Williams, a clock- and watchmaker of Llancarfan, presumably a relative of the Henry Williams mentioned above. Subsequent generations of the Williams family continued the watch- and clockmaking traditions in Llancarfan until the middle of the nineteenth century.[3]

As is clear from surviving examples of their work, Richard and Thomas Watkin were active clockmakers in Merthyr Tydfil in the latter half of the eighteenth century, as were Charles Vaughan in Pontypool, and Griffith Williams in Newport.

During the eighteenth century there were at least as many people engaged in making clocks and watches, either on a full- or part-time basis, in Vale villages such as Llancarfan, Wenvoe, Laleston and St Mellons, and in communities such as Llantrisant and Merthyr Tydfil, as in Cardiff itself. These clockmakers were also general craftsmen, able to repair a gun for a neighbour or mend a lock or a coffee-pot for the gentry.

Although directories are not necessarily complete or always accurate, they give a good indication of commercial trends. Directory entries for selected years from 1795 to 1889 show the mushrooming increase in the numbers of clock- and watch-makers in Cardiff as the town grew, the industrialisation of Glamorgan gathered pace, and the emphasis in the clock and watch trade changed from primary

---

[1] Ibid. p. 103.

[2] *Cardiff Records*, III–IV, passim.

[3] Glamorgan Record Office, Llancarfan Parish Register; *Morgannwg*, XXXI (1987), p. 20.

production by skilled craftsmen to repairing and retailing.[1]

The *Universal British Directory* (*c.* 1795) lists two watchmakers in Merthyr Tydfil (Owen Bowen and John Trick), one in Bridgend (Thomas Bowen), one in Caerphilly (Lewis Stradling) and two in Cardiff, namely John Thackwell and W. Wilson. The *Cardiff Directory* (1796) also mentions Thackwell and Wilson as watch- and clockmakers in Cardiff. The *Complete Directory* (1813) also lists only two: Thackwell and Michael Marks. Thereafter the number slowly increased: *Bird's Directory* (1829) and *Pigot's Directory* (1835) both list the same four watch- and clockmakers in Cardiff, Mark Marks, Nicholas Schriber, Solomon Marks and John Thackwell. In 1830 *Pigot's Directory* includes three clockmakers in Newport (William & Edward Frost, Thomas Street; William Latch, High Street; and Evan Williams, Westgate Street), two in Pontypool (Samuel Shill and John Evans), and John Harvey in Abersychan, near Pontypool.

The published Census reports show the growth of population:[2]

Year	Cardiff	Merthyr Tydfil	Newport
1801	1,870	7,705	1,135
1821	3,521	17,404	4,000
1841	10,077	34,977	10,492
1861	32,954	49,794	23,249
1881	82,761	48,861	38,469
1901	164,333	69,228	67,270

Not until 1881 did Cardiff overtake Merthyr Tydfil as the most populous town in Wales.

Clockmakers' premises were generally located in the central business district of towns: High Street in Merthyr Tydfil, and Angel Street, High Street, Duke Street and St Mary Street in Cardiff.

Shortly before the middle of the century, according to *Pigot's Directory* (1844) there were six clockmakers in Cardiff, and the same number in Merthyr Tydfil:

Cardiff 1844	**Merthyr Tydfil 1844**
James Trotter Barry, 6 Duke St	Lewis Beynon, High Street
John Evans, 99 St Mary St	William James, Rhymney
Mark Marks, 9 St Mary St	Jenkin Thomas Jenkins, Dowlais

---

[1] For full details of directories mentioned in the following paragraphs see H. Llewellyn, *A bibliography of Cardiff directories, 1795–1978* (Cardiff, 1990).

[2] Williams *Digest of Welsh Historical Statistics*, pp. 62–7.

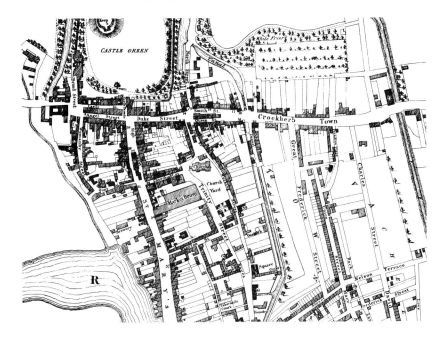

Fig. 5. *Above:* Extract from John Wood's map of Cardiff (*c.* 1840). Most of the early clockmakers had their premises in Angel Street, High Street, Duke Street and St Mary's Street *(Glamorgan Archive Service)*. *Below left:* The England-Träger: a couple of German clock-pedlars from the Black Forest, from a postcard published at Lübeck, posted in 1960. *Below right:* A typical advertisement by Wladyslaw Spiridion (*Bird's Cardiff Directory*, 1858).

Solomon Marks, 2 St Mary St
John Thackwell, 12 Angel St
George Truscott, Bridge St
David Jones, High Street
William Jones, High Street & Tredegar
Andrew Pfaff & Co. (German clocks)

Pigot's Directory also lists two clockmakers each in Bridgend (Thomas David and William Griffiths), Cowbridge (Samuel Marks and George Morgan), and Newbridge (i.e. Pontypridd) (Evan Davies and David Goodman).

Five years later, *Hunt's Directory* (1849) listed seven watch- and clockmakers in Pontypool and nine in Merthyr Tydfil, as against only seven in Cardiff. Joanna Greenlaw has also shown that until the middle of the nineteenth century Swansea was a more important commercial centre and had more watch- and clockmakers than Cardiff.[1]

Ewen's Cardiff Directory (1855) lists twelve makers:

* James Trotter Barry, 6 Duke St
* John Collings, 19 Bute St
  Michael Furtwangler, 72 Frederick St
* Henry Grant, 29 Duke St
* James Oliver Ingram, 29 High St
  Levy Marks, James St
Solomon Marks & Co, 101 Bute St
Raymond Heitzman, 82 St Mary St
Benjamin Knight, 277 Bute Street
E. Phillips, 295 Bute St
William Samuel, 36 St Mary St
Charles Thackwell, 12 Angel St

* indicates that they were also silversmiths and jewellers

Two directories published in 1863 give slightly different numbers for Cardiff: *Wakeford's Directory* lists fifteen, whereas *Duncan & Ward's Directory* lists the following eighteen:

Barry, J. T., 8 Duke St
Collins, Samuel, 9 Bute St
Collings, S., 219 Bute St
Collins, J., 47 Bute St
Furtwangler, M., 231 Bute Rd
Heitzman, R., 29 St Mary St
Ingram, J.J., 32 High St
Marks & Son, 101 Rothsay Terrace
Matt & Co., 17 Trinity St
Moretti, R., 265 Bute St
Nolcini, J., 18 Bridge St
Pedrazzini, B., 18 High St
Primavesi & Son, James St
Roberts, John, 19 Bute St
Samuel, Moses, 22 Bridge St
Spiridion, W., 24 Duke St
Thackwell, Miss, 13 Angel St
Weichert, W., 112 Bute St

By the 1880s *Butcher's Directory* (1882–3) lists no fewer than 29 makers in Cardiff:

---

[1] J. Greenlaw, *Swansea Clocks* (1997), p. 17.

Jas. T. Barry	Jno Ingram	Chas. Pring
G. Best	Hyam Jacobs	Giovanni Solari
I. Bleibel	Elias Kaiser	W. Spiridion
Edward Bonfiglio	Richard Kaiser	Tainsh Bros.
Mrs S. Bradford	A. Kaltenbach	Hy J. Thatcher
M.R. Collings	E. Kaltenbach	T.H. Ward
Albert Halter	William Long	R.P. Watson
William Harris	Solomon Marks	Wm Weichert
R. Heitzman	Rocco Moretti	W.H. Winstone
Julius Hettich	William Morris	

By 1891 the population of Cardiff had reached 116,207. Between 1801 and 1911 Cardiff's population increased nearly a hundredfold, to 182,259. The growth in the watch and clock business in Cardiff clearly mirrored the development of the Glamorgan coalfield, and the growth in population and in coal exports from the port during the nineteenth century. By the end of the century, Pontypridd and district boasted more watch- and clockmakers than Merthyr Tydfil, and nearly as many as Cardiff itself. Aberdare, Barry, Newport and Bridgend were also important centres for the trade, which by this date had virtually disappeared from the villages of the Vale of Glamorgan, and even from the smaller towns such as Cowbridge.

One thing that emerges clearly from any scrutiny of the lists of watch- and clockmakers is the increasing preponderance of foreign names during the nineteenth century. Even before 1800, the diarist William Thomas had drawn attention to the English origins of some of the earliest clockmakers in Glamorgan—men like Henry Williams of Llancarfan, who hailed from Gloucestershire, the Phillips brothers of Pentyrch and Llantrisant, and Samuel Verrier (Varrier) who was 'an Englishman by birth', from a family of Huguenots. Several of the immigrant clockmakers came to South Wales from Gloucestershire, and Bristol was an important early centre of the clockmaking trade, its influence extending on both sides of the Channel in the south-west of England and in South Wales.

Political instability, rural poverty, religious persecution, and social and commercial pressures caused increasing numbers of refugees, often skilled craftsmen, to emigrate to Britain. They came from France, the Low Countries, Poland, Russia and Italy, but by far the greatest influx was from Germany, and especially from Baden, in the south-west which includes the Black Forest region. In 1806 the Grand Duchy of Baden took over extensive but impoverished areas of north Baden. Fragmented landownership, feudal conditions of servitude and rural poverty induced many Baden people, including clockmakers, to seek new opportunities abroad, and in fact the state of Baden actively promoted emigration in the early part of the nineteenth century.[1]

During the eighteenth century the Black Forest area of Germany had already developed into a major centre of clockmaking. One characteristic feature of Black Forest clockmaking was the extensive and skilled use of wood, a raw material that was cheaper and more readily available than brass. Wood was used not merely for the clock cases, which were often ornately carved, and for the dials, which were usually brightly painted, but even the plates, pillars, wheels and other parts of the clock movements were fashioned from wood instead of brass. German clocks with partly wooden movements are sometimes encountered in South Wales. For example, a cheap longcase clock of the mid nineteenth century, signed 'Ingram, Abergavenny', seen recently in Pontypridd, had a convex painted wooden dial, while its movement had wooden plates and pillars instead of the conventional brass.

Another characteristic feature of Black Forest clocks was the ingenious use of moving figures or automata. In 1796 Franz Steyrer listed nine characteristic types of moving parts or automata for clocks, usually operating on the hour:[2]

> sun, moon or planets moving
> a Franciscan friar ringing a bell
> a sentry saluting
> a butcher killing an ox
> two billy goats butting each other
> bears dancing
> a cuckoo (or a quail) calling
> a grinder sharpening a knife or scissors
> twelve apostles or other figures striking the quarters and hours

Several of these types of moving figures appear in clocks in South Wales. Moving moons, cuckoos, and Adam-and-Eve figures are quite common. One good example of an eight-day longcase clock by W.

---

[1] *Die regionalen ländlichen Freilichtmuseen in Baden-Württemberg* ('The regional rural open-air museums of Baden-Württemberg') (Wolfegg, 1997).

[2] F. Steyrer, *Geschichte der Schwarzwälder Uhrenmacherkunst* ('History of the Black Forest Craft of Clockmaking') (Freyburg, 1796), p. 21.

Ansell of Pontypool has the figure of a butcher in the arch who, on each hour, strikes an ox in the neck with a large knife the appropriate number of times. Understandably enough, rocking ships, so popular a feature of Welsh clocks, did not feature among the favourite types in the Black Forest, an inland area far from the sea.

The standard work on the Black Forest clockmakers lists the long-established clock-making families of the Black Forest, among which are many names that had become familiar in South Wales during the nineteenth century: Beha, Dold, Duffner, Furtwangler, Furtwengler, Ganz, Grieshuber, Heitzmann, Kaiser, Kaltenbach, Kern, Pfaff and Winterhalder.[1] Initially the Black Foresters came to Britain as *England-Träger*, travelling salesmen or pedlars, carrying clocks around on their backs for sale. Soon, the more enterprising among them were tempted to set up their own clock businesses in the major centres in Britain. As early as 1821, Franz Simon Meyer could state:

> the Black Forest clock-business is very widespread in England. There are Neustadt clock-dealers in Manchester, Birmingham and all the important towns in the island. But in London there are hundreds of them. In nearly every street you see our good Black Foresters, and despite their ignorance of the English language the industrious among them nearly always do good business. Many of these pedlars buy in from Kleyser ...[2]

In 1843 an official German Commission, set up to inquire into the clock business, reported that, of all the pedlars from the Black Forest known to be trading in various countries throughout the world, over 40 per cent were in Great Britain (297 out of 730).[3]

The rapidly expanding industrial area of South Wales, with its high wages and growing demand for consumer goods, acted as a magnet to these Black Forest clock-dealers, many of whom set themselves up in business here, often making use of networks of family contacts. While getting established, some of the young men even slept in the shop to save money, and although initially their ambition might often have been to save up enough to buy a substantial property in the

---

[1] Gert Bender, *Die Uhrenmacher des hohen Schwarzwaldes und ihre Werke* ('The Clockmakers of the High Black Forest and their Works' (Villingen, 1975–8).

[2] Ibid., II, 288.

[3] Ibid., II, 300.

Black Forest to retire to, in the natural course of events many settled permanently in South Wales, where their descendants remain.

Of the nine watch- and clockmakers listed in Merthyr Tydfil in *Hunt's Directory* (1849), three dealt only in German clocks, namely Thomas Dold, Andrew Pfaff and Romanus Pfaff, all of whom had premises in High Street.

In his study of clockmaking in Glamorgan in the eighteenth and nineteenth centuries, Peter Jarvis identifies 49 European immigrants (watchmakers, clockmakers, chronometer-makers and jewellers) from the census enumerators' books of 1851 and 1881. Of these, 44 were born in Germany, the majority of them in Baden (Black Forest); three were from Poland, and just one each from Austria and Russia. Some of these immigrants applied for naturalisation, for example the Cardiff clockmaker Raymond Heitzman (19 April 1859), his brother Pius Heitzman (8 March 1871) and Joseph Rombach (11 January 1855).[1]

It is these foreigners whose names have caused such problems of spelling to the compilers of local directories and to historians of horology, not to mention modern restorers of old clock dials. To give just two examples, F. Schönhardt of Dowlais, usually given as 'Schonhardt' in directories, was also written, correctly, as Schoenhardt, but his name appears as Schornhort on the otherwise nicely restored dial of a large longcase clock seen recently for sale in Ross-on-Wye. Nicholas Shriber of Cardiff is recorded in directories and parish registers with various spellings: Scribor, Scriber, Shriber and Schriber; Peate gives his name as Nicholas Schriebner, and the correct modern spelling (Schreiber) has been seen on restored clock dials.

The cosmopolitan nature of late nineteenth-century Cardiff and the trade is graphically demonstrated by Weichert's trade card in five languages, English, Italian, French, German, and Greek.

Mass-production techniques brought timepieces within the reach of most members of society. Even the humblest household needed a clock, and every working man aspired to a watch. The local watch- and clock-makers of Cardiff and the valleys catered for this mass market and of course, wherever possible, they also sold to the wealthier members of society. For example, various Cardiff clockmakers supplied clocks for the servants' quarters at Cardiff Castle

---

[1] P.H. Jarvis 'Clockmaking in Glamorgan in the eighteenth and nineteenth centuries' (Dissertation, Diploma in Continuing Education (Local History), University of Wales, Cardiff, 1995), p. 13.

during the nineteenth century. An inventory of the Castle in 1931 lists three clocks: in the pantry stood a mahogany longcase clock with a painted dial by M. Marks (valued at £12 10s. 0d.); in the housemaids' sitting room hung a round wall clock by Jos. Collings (£4 10s. 0d.); and in the servants' hall stood another and more superior mahogany longcase eight-day clock with white enamelled dial by Ingram of Cardiff (£25).[1]

Few fortunes were made in the local watch and clock business. In the early years for many men it was a part-time business combined with some other quite unrelated activity. For example, Alexander Wilson was a customs surveyor at Cardiff, Henry Williams was a farmer at Llancarfan, and George Morgan was the landlord of the Bear Hotel, Cowbridge. Many clockmakers prospered, but some were brought down by failing eyesight, debt and drink. Moreover, competition in the retail trade was particularly keen in the second half of the nineteenth century.

A few of the early clockmakers have achieved lasting fame by the outstanding quality of their surviving work, which was recognised by wealthy local patrons. Such men included Henry Williams of Llancarfan, Griffith Williams of Newport, and Charles Vaughan of Pontypool, all of whom produced clocks which graced great houses such as Itton Court, Llanvair Grange and Tredegar House, all in Monmouthshire.

Some clockmakers established businesses which remained within their family over several generations. One good example is the Kaltenbach business at Caroline Street, Cardiff, which flourished from the 1860s until its closure in 1998.

A few clockmakers stand out as great individual characters in local society. One such was John Thackwell (1769–1845) of Cardiff who became an alderman and bailiff. He first repaired and later replaced the clock at St John's Church, Cardiff. He and his family were prolific makers, and many of his longcase and wall clocks survive locally. He had his own personalised puzzle jug, coloured yellow, pink and chocolate, and inscribed 'SUCCESS TO THE CLOCK AND WATCH TRADE. J. E. T. CARDIFF 1880'. This inscription is in Thomas Pardoe's own hand, and the date is either an error or, more likely, a joke on the clockmakers; it must have been 1808. Clock-

---

[1] Cardiff Castle Collections, Unnumbered MS.

makers called Thackwell, presumably relatives, were also active in Gloucester during the period *c.* 1760–1840.[1]

Charles Vaughan of Pontypool was an important and quite prolific clockmaker between about 1735 and 1795. Nothing definite is known of his family antecedents, and he cannot be traced among the various Vaughans in the parish registers of Trevethin, Panteg or Llanvihangel Pontymoile. However, his surviving brass-dial longcase clocks, mostly thirty-hour but including some better eight-day clocks, show him to have been a good craftsman. He signed his clocks 'Vaughan', 'C. Vaughan', 'Cha. Vaughan' or 'CHARLES VAUGHAN', and the place-name is variously inscribed as 'Pont-Pwll', 'Pont Pool', 'Pont pool', 'Pontpool', or merely 'P.P.'. Some of his dials also bear a date, e.g. 1759, 1760, 1777 and 1796. The cases of his cheaper clocks were in plain oak, or sometimes elm, some with simple gilded wooden finials and an open-backed flat box on top of the flat hood. The case of one of his eight-day clocks, formerly at Itton Court, was in black carved oak. On the movements of his cheaper thirty-hour clocks he usually saved brass by using iron of rectangular cross-section for the bottom pillars (pierced and threaded for the seatboard screws), and sometimes also for the top pillars. The front and back plates of his thirty-hour movements characteristically have large pieces cut out from the bottom or the corners or both, in order to save brass. A thirty-hour clock, now in New York, has moonwork in the arch, and 'TIME IS VALUABLE' engraved on the globes.

Another famous character, more noted locally as a bard, author of druidical works and self-styled archdruid rather than as a clockmaker, was Evan Davies (1801–88) of Pontypridd, better known under his bardic name Ieuan Myvyr or Myvyr Morganwg.[2]

David Jones of Merthyr Tydfil was another noted clockmaker. He was born in 1760 near Llangadog, and taught himself the art of clockmaking. In 1816 he moved to Merthyr Tydfil where 'with little capital beyond his skill' he opened a shop opposite the field where the Market House was later built, soon establishing an excellent business. About 1828 he invented and exhibited a fine illuminated clock called 'a night watch or regulator', the only one in Wales, and

---

[1] *Cardiff Records*, VI (index sv. Thackwell); E.M. Nance, *The pottery and porcelain of Swansea and Nantgarw* (1942), p. 87; G. Dowler, *Clockmakers and watchmakers of Gloucestershire* (1984), sv. Thackwell.

[2] *Dictionary of Welsh Biography* (1959), p. 123.

with only two others in existence, one in Scotland the other in London. At eleven o'clock the mechanism was so arranged as to put the light out, to show that people should be at home at that time. He repaired the celebrated clock belonging to Williams Perthygleison (on Mynydd Merthyr): 'a masterpiece, with its chimes and bells, but of most intricate construction'. While president of the Cymmrodorion White Horse Society he invented the 'celebrated clock watch which, in addition to the usual properties of the repeater, spontaneously struck the hours without touching the pendant or anything whatever ... The invention of our old townsman was not only different, but in many respects superior to those of Quare and Barlow ... Jones's instruments, tools, and all materials of the clock were made *by himself*. In 1837 the White Horse Society offered a prize of £1 for six englynion to 'Y Gloch Oriawr'. The prize was won by William Evans (Cawr Cynon), the first and last of his winning englynion being as follows:

> Oriawr gain gywrain yn gwirio—i'r byd
>   Fawr wybodaeth Cymro
>   Gwn i'r Sais ddweyd na roes o
>   Un i atteb hon etto.
> Dyfais o eiddo Dafydd—Jones ydyw,
>   Hen sywdog Oriorydd
>   Er arwydd ar hwyr ei ddydd
>   I'w fro gain o'i fawr gynnydd.

Of venerable appearance and affable character, in his later years David Jones was familiarly known as 'Yr Hen Llwyd'. He died in 1842.[1]

J.D. Willams (James David Williams) was a watchmaker, goldsmith, silversmith and jeweller in Merthyr Tydfil for nearly fifty years, from 1856. He sold the typical range of clocks and watches, but is best known for the fine regulator nearly seven feet high and two feet wide at the base, which stood on display in the shop window in Merthyr for over a hundred years. After the death of the founder, the business was continued by his son as J. D. Williams & Co. In 1904 new premises were acquired on the corner of High Street and

---

[1] *Yr Haul*, 1842, p. 291; *Merthyr Telegraph*, 9 Feb. 1861, p. 4.

Glebeland Street, with a portrait of the late J. D. Williams in a massive mahogany case under a handsome clock. The 'town clock' on top of the building had a dial five feet in diameter, illuminated at night.[1]

Wladyslaw Spiridion (originally Wladyslaw Spiridion Kliszczewski) was a refugee Pole who left home after the Russo-Polish War of 1830–1, was imprisoned at Trieste as a political prisoner, and eventually reached London in 1838, where he was apprenticed to a watchmaker. In 1844 he was employed in Cardiff by Henry Grant, whose sister he later married. In 1855 he bought Grant's business. He died in 1891, and his son Joseph continued the business until 1920.[2] The firm of Spiridion & Son (2a Duke Street, Cardiff) erected the clock at the Cardiff Fine Art, Industrial & Maritime Exhibition of 1896: 'The dial is ten feet in diameter, painted black, with raised gold figures and gold hands ... The firm also made and fitted up in the turret of the Old Town Hall in Old Cardiff a clock which is in appearance, as near as possible, a facsimile of the original'.[3] The turret clock of the former St David's Hospital in Cowbridge Road, Cardiff, rebuilt in its existing form by the poor law union in the 1880s, has 'SPIRIDION CARDIFF' on the dial. In addition to the normal wide range of clocks and watches, the firm also produced large table clocks in ornate Gothic fretwork cases, designed and made the Cardiff mayoral chain and badge, repaired maces, and valued the Corporation regalia. In 1878 Spiridion advertised 'guaranteed patent English lever watches, capped and jewelled, gold balance, sunk seconds, in strong silver case £3 17s. 6d.'

Towards the end of the nineteenth century, with the burgeoning of civic pride and commercial confidence based on the thriving coal export trade, several very imposing four-dial turret clocks were erected in Cardiff. Most were manufactured in England, but were wound and maintained by local firms such as Spiridion & Son. Cardiff Castle tower clock, made in 1870 by E. Dent & Co. of London, is described below. The clock tower of the terracotta brick Pierhead Building in the Docks, designed by William Frame, a pupil and assistant of William Burges, was opened in 1897; its original

---

[1] *Merthyr Express*, 3 Sept. 1904, p. 7.

[2] Peate, *Watch and clock makers*, p. 77.

[3] *Cardiff Fine Art, Industrial & Maritime Exhibition Catalogue* (1896), p. 231.

turret clock was made by William Potts & Son of Leeds.[1] The turret clock of the City Hall in Cathays Park, made by Gillett & Johnston of West Croydon, was formally started by the mayor and mayoress of Cardiff on St David's Day 1905, some eight months before Cardiff became a city and twenty months before the City Hall was officially opened. The firm of Spiridion & Son, recommended by Gillett & Johnston to wind and maintain the clock, explained the work involved in a letter to the Council in 1905:

> I have carefully gone into the matter of winding and maintaining the above clock ... the actual time consumed in winding, not to mention going to the top of the tower and back, will be three hours per week for two men; as it is physically impossible for one to wind the three weights without assistance. Then comes the care of the clock and chimes ... which cannot be done except by a man experienced in such work. I, therefore, offer to take full responsibility, including cleaning, winding, maintaining, etc., for twelve shillings and sixpence per week. Unforeseen and unavoidable accidents, wear and tear of lines, pulleys, cranks, etc., excepted.[2]

During the second half of the nineteenth century most watch- and clockmakers were concentrating on retailing items manufactured elsewhere and diversifying their range of stock and services. Traditional longcase clocks were fast losing their popularity, being supplanted by smaller and cheaper wall, mantel, bracket and carriage clocks.

It is fortunate, therefore, that we have a comprehensive illustrated price list of the wide range of stock available from a typical watch- and clock-maker in South Wales at the end of the nineteenth century. Surprisingly, this emanates not from one of the leading or old-established businesses in Cardiff or Swansea, but from the small coalmining community of Blaengarw, in central Glamorgan, which in the late nineteenth century grew rapidly from a sparsely populated rural community into a thriving industrial settlement of several thousand, based on three collieries. In April 1891 it was said of the

---

[1] Information kindly supplied by ABP, Cardiff.

[2] *City Hall Clock Tower* (Leaflet issued by Cardiff City Council, 1986).

town that: 'Coal-dust cannot as yet be said to have set its mark on the place, and there is consequently a brand-new air about the streets and buildings'.[1] By 1910 Blaengarw boasted one church, five chapels, a Salvation Army barracks, a Working Men's Hall & Institute, three branch banks, two elementary schools with room for six hundred children and four hundred infants, and a busy little commercial centre which included one watch and clock business owned by Howell Morgan.[2]

Advertising himself as 'Watch and Clock Maker, Jeweller &c.', Howell Morgan ran his business at Park House, The Strand, Blaengarw, from the 1890s to 1926. Morgan, an enterprising businessman, produced a detailed price list, which was printed by Crick & Co., High Cross Works, Tottenham. This lavishly illustrated catalogue is undated, but internal evidence indicates a date of about 1900. It is a uniquely valuable document, showing, as it does, the whole range of watches, clocks and many other items sold by a typical South Wales trader at the start of the Edwardian period. Such illustrated catalogues of clocks and watches are extremely rare, and no other example is known from South Wales. For this reason, the entire booklet has been reproduced here as an Appendix.

---

[1] *Central Glamorgan Gazette*, 3 April 1891.
[2] *Kelly's Directory of South Wales* (1910), pp. 693–5.

Features of Local Longcase Clocks

> My grandfather's clock was too large for the shelf,
> So it stood ninety years on the floor:
> It was taller by half than the old man himself,
> Though it weighed not a pennyweight more.
>   (*Grandfather's Clock;* words and music by Henry C. Work, 1875)

THIS CHAPTER IS DEVOTED MAINLY to the longcase clock, which is currently enjoying a period of renewed popularity. The increasing interest in these grand old survivors from earlier centuries is seen from the number of published articles and books on them and, of course, the steadily rising prices which they command in salerooms and auctions. This is not a comprehensive account, and merely points out some particular features of local longcase clocks within the general context.

Cases

The development of clock case design in England and Scotland has been described in detail by Brian Loomes and in Wales very briefly by Iorwerth Peate.[1] In South Wales, case styles tended to follow those of southern England, whereas North Wales was more influenced by styles popular in the northern counties of Cheshire, Lancashire and Yorkshire.

Throughout Wales, case-makers in general were conservative, frugal and unostentatious in their use of materials—our local clocks are rarely found in the more florid or expensive types of cases. Early cases tended to be relatively slender, flat-topped with a long narrow door to the trunk, and a lift-up hood. Oak was commonly used, indeed almost universally, up until the end of the eighteenth century,

---

[1] B. Loomes, *Grandfather clocks and their cases* (1989); Peate, *Watch and clock makers*.

and cases were generally made of local oak by carpenters in close contact with the actual clockmaker. Inlay was sometimes used on the cases, but elaborate marquetry or parquetry was very much the exception. Very few walnut cases or painted and lacquered Chinese-style cases are to be found in South Wales, though Charles Vaughan and William Evans (Pontypool) produced some clocks in 'japanned', i.e. lacquered, cases in the eighteenth century.

In the second half of the eighteenth century, mahogany became more generally available, via the port of Bristol, and clock cases followed furniture styles, the use of mahogany veneer on stained softwood carcasses becoming more common and remaining popular throughout the nineteenth century. Clock cases became broader, with a shorter trunk door, and swan-necked pediments and slide-forward hoods, often domed to accommodate arches or moon-work.

As well as the more fashionable mahogany, cases also were made of plain softwood, stained or painted, especially for the smaller and cheaper thirty-hour clocks. The survival rate of these cheap softwood cases was significantly lower than that of the up-market mahogany cases, and indeed their solid oak predecessors.

Occasionally elm was used as a cheaper alternative to oak. Wood of alder—locally available, easy to work and to stain—has also been identified in cases of a few clocks from rural parts of South Wales. Cases made of fruit woods (apple, pear, cherry) are occasionally found, for example the case of a clock by one of the Williams clockmakers of Llancarfan in the Vale of Glamorgan. Yew, too, was locally available and sometimes used by furniture makers but rarely, if ever, for clock cases.

Though Welsh oak furniture traditionally was often lavishly decorated by carving, oak clock cases very rarely bear any carved decoration, either as an original feature or as a later (Victorian) embellishment. The patriotic 'Three Feathers' motif is sometimes found inlaid in the trunk door of Welsh clocks in the nineteenth century, for example a clock by William Williams of Merthyr Tydfil, *c.* 1825. However, the 'Three Feathers' is not exclusive to Welsh clocks, and is also seen for example in cases by Cheshire makers.

There are several stylistic features which distinguish many of the mahogany-cased clocks which became so popular in South Wales for much of the nineteenth century. These are: the notched crest or wavy-edged cresting piece of the domed hood, often with two parallel semi-circular strings inlaid below it; the wavy or scalloped border surrounding the glass of the hood door; and the rope-twist hood pillars with brass caps and bases, sometimes of ornate Corinthian style.

Though these stylistic features are characteristic of many cases in all parts of South Wales, they are not exclusively Welsh, being also seen, for example, in longcase clocks in many parts of the West of England, such as Ledbury, Bath, Bristol, Frome and the coastal towns of Somerset and North Devon. This seems to indicate a distribution area in the counties on both sides of the Bristol Channel and radiating out from Bristol itself. In the eighteenth and early nineteenth centuries Bristol was a major port, the trading and population centre of a large region and, with Bath, the fashion-setter for the West Country and South Wales. Regular coastal trade from Bristol to the little harbours on the Welsh and English shores of the Severn could supply exotic imported and manufactured goods of all kinds. Delivery of clock cases by sea would have presented no difficulty before the days of regular and reliable transport by road and rail.

It has not been possible to discover which case-makers were responsible for these distinctive stylistic features. Case-makers rarely put their labels or names on cases, and other documentary evidence is sadly lacking for Wales. The large numbers of similar or identical cases seen for Swansea, Merthyr and Cardiff clocks (not to mention Bristol and West of England clocks) strongly suggest one centre of case design and probably manufacture.

One of the most famous and prolific cabinet-makers and case-makers in Bristol was William Cock (*fl.* 1816–40), who operated at Hillgrove Street, Bush Street and Lower Cheltenham Street. His trade-card states: 'clock-cases made in the neatest manner and on moderate terms', but there is no direct evidence that he was the man responsible for manufacturing these distinctive mahogany cases. Cabinet-makers in Swansea and Cardiff certainly had the facilities to make fine veneered cases to standard designs, but there is no absolute proof that they did, and as the dials and movements were ordered from Birmingham, so the cases too could have been ordered from Bristol.

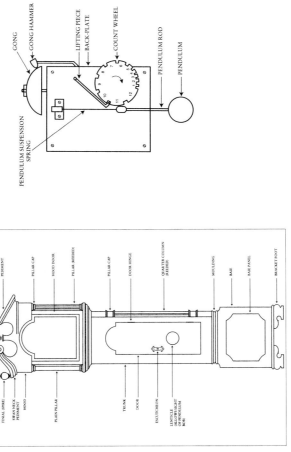
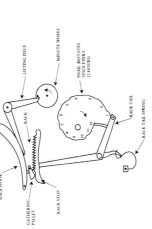

Fig. 6. *Centre*: Terminology of the parts of a clock case. *Left*: Rack-striking system. *Right*: Count-wheel striking mechanism (Joanna Greenlaw, *Swansea Clocks*).

Movements

Until the 1650s, lantern or bracket clocks, with crown wheel, verge escapement, foliot balance and weights, were the standard domestic clock in Britain. Despite their inferior timekeeping, they had remained essentially unchanged from late medieval times. No extant examples or even pictorial representations are known from South Wales.

Several important technical innovations in the second half of the seventeenth century provided much greater accuracy in timekeeping and led directly to the development of the longcase clock. First, Christiaan Huygens in Holland elucidated the application of the pendulum. This was quickly introduced to London by Ahasuerus Fromanteel in 1658. Then the anchor escapement was invented in the 1670s either by William Clement or Dr Robert Hooke. In 1676 the Revd Edward Barlow invented the rack striking mechanism.

As a result of all this innovative activity, clock dials and movements, initially similar to those of lantern clocks, could be enclosed in a glass-fronted hood, while the more accurate long pendulum (39¼ inches, giving a one-second beat with a small arc of swing) and the weight(s) could be conveniently accommodated and protected inside a long wooden case below, and a seconds dial could also be incorporated. This arrangement was both technically successful and aesthetically pleasing, and indeed the longcase remained the most popular domestic clock in Britain for over two hundred years.

Although lantern clocks were being produced by several clockmakers in Bristol, north Somerset and Gloucestershire from the 1640s onwards, it is significant that no examples of lantern clocks are known from any maker in South Wales. This indicates that South Wales—rural, remote, thinly populated and comparatively poor—could not really support local clockmakers until the longcase clock started to become popular at the end of the seventeenth century. Before that, the few wealthy individuals in South Wales must have acquired their domestic clocks from England.

Two main types of longcase clocks were developed: the thirty-hour birdcage clock, and the eight-day clock. The thirty-hour birdcage or posted movement was a direct development of the wall-hung lantern clock, with upright pillars in the form of a 'cage' which were pierced to carry the pivots of the arbors. In South Wales, some thirty-hour birdcage movements were still being made by men such as Phillips (Llantrisant), Edmond Edmonds (Laleston) and Charles Vaughan

(Pontypool) in the middle of the eighteenth century, though towards the end of the century and always thereafter thirty-hour clocks had plated movements. Indeed, some of Charles Vaughan's plated movements represented a transitional stage from the earlier birdcage or posted movement—to save brass he cut away so much from his plates that they resemble 'semi-posts'. Thirty-hour clocks are wound by pulling the endless rope or chain to raise the single weight. The eight-day clocks were naturally more expensive, and had the going and striking trains of the movement contained between two parallel vertical brass plates. The dials of eight-day clocks have holes for the winding 'squares' which are wound up by key.

The striking mechanisms of longcase clocks are of two kinds, namely the count-wheel system or the rack system. The count-wheel system, also known as locking-wheel or locking-plate striking, was the first and was used on all clocks, both eight-day and thirty-hour. It continued to be used, especially on the cheaper thirty-hour clocks, up until the end of the nineteenth century. The rack-striking system, invented in 1676, was slower to gain acceptance but eventually became the normal form for most eight-day clocks from the mid eighteenth century until the final demise of longcase clock manufacture towards the end of the nineteenth century.

Clock weights were usually iron, cast to order in a local foundry. Weights commonly ranged from seven pounds for thirty-hour clocks to fourteen pounds for eight-day clocks, and sometimes the initials of the individual clockmaker and/or the weight in pounds were included in the casting to appear on the actual weights. For example, for his thirty-hour clocks Charles Vaughan of Pontypool used a cast weight of 9lb or 10lb, bearing the figure 9 or 10 on the top or bottom of the weight. Iron weights cast for Thomas Davies of Monmouth bore the initials $^T_M{}^D$. Occasionally an emblem, e.g. a dragon, was also included on cast clock weights in South Wales.

The advent of painted (japanned) dials towards the end of the eighteenth century led dialmakers to use a falseplate of cast or sheet iron, in order to facilitate attachment of the dial to the front plate of the movement. Falseplates often bear the name of the dialmaker, cast or stamped on the plate. Most of the dials encountered in South Wales were manufactured in Birmingham, where many different firms were engaged in the trade. Among the names encountered most frequently on the falseplates of South Wales clocks are Wilson, Walker, Osborne, Finnemore, and Wilkes, all of Birmingham.

LOCAL LONGCASE CLOCKS 45

Fig. 7. *Above left:* Movement of thirty-hour clock by Charles Vaughan (Pontypool), showing characteristic cut-away bottom corner of front plate, and bottom pillar of iron, rectangular in cross-section *(Museum of Welsh Life)*. *Above right:* Plate (6 × 4 inches) showing the pieces cut away to save brass. *Below:* Brass dial of clock by Richard Watkin, Merthyr Tydfil; the chapter ring has full minute and quarter-hour bands. (Peate, *Clock and watchmakers*).

Other Birmingham manufacturers' names have been recorded on falseplates in South Wales, as well as a couple of names that cannot be positively identified and located. Perhaps surprisingly, it appears that no painted dials and falseplates were ever manufactured in South Wales itself, even though the necessary skills and experience were certainly available in several places, most notably Pontypool.

Some eighteenth-century movements show idiosyncrasies of design—for example, scribings on the plates where the clockmaker had laid out his trains of wheels. Charles Vaughan (Pontypool) used iron pillars of rectangular cross-section instead of the conventional round brass ones in his thirty-hour clocks, and he also cut pieces out of the plates, presumably in order to save brass. However, by the nineteenth century the movements of South Wales clocks show no technical features that distinguish them in any way from their English counterparts. Only very occasionally are numbers or initials found stamped or scratched on a plate, the significance of which is now lost.

Dials

The stylistic features of brass dials of longcase clocks have been ably described by Brian Loomes,[1] while development of white or painted dials has been described in great detail both by Loomes and also Mrs M.F. Tennant.[2]

Early brass-dial clocks were often made with only a single (hour) hand, and in these clocks the chapter ring is divided only into quarter-hours, for example clocks by Thomas Shinn of 'Mathan' (i.e. Mathern, near Chepstow) and Edmond Edmonds of Laleston in Glamorgan. Clocks with both hour and minute hands usually had a chapter ring with an inner band showing the quarter-hours, and an outer band divided into minutes, for example the eight-day clock by Richard Watkin (Merthyr Tydfil), illustrated by Peate.[3] Individual makers, e.g. Charles Vaughan of Pontypool, spanned this period of change: the earliest of his clocks have a single hand and a quarter-hour band, while his later two-handed clocks have a chapter ring with

---

[1] e.g. *Grandfather Clocks and their Cases* (1989); *Brass Dial Clocks* (1998).

[2] Loomes, *White Dial Clocks* (1989); M.F. Tennant, *Longcase Painted Dials* (1995).

[3] Peate, *Watch and clock makers* (1975 ed.), plate IV.

a full minute band and a quarter-hour band. As two-handed clocks became the norm, generally only the minute band was retained. Early brass dials were small and square, typically nine inches. The size of square dials generally increased during the eighteenth century, with arches and sometimes moon-work or automata on better (usually eight-day) clocks. High-quality engraved decoration is seen on some brass dials, even on the cheaper thirty-hour clocks, for example work by Griffith Williams of Newport and David Phillip of Llantrisant.

Some of the earlier local makers numbered their clocks. Thomas Shinn of Mathern was unusual in engraving the number on the face of his brass dials. One of his clocks illustrated by Loomes bears the number 246 prominently engraved in the centre of the dial.[1] A few clocks, e.g. some of Charles Vaughan's, have the year of manufacture engraved on the front of the dial.

Towards the end of the eighteenth century, brass dials were quite rapidly and generally supplanted by the new and highly popular painted dials. As the white or painted dials were decorated by a relatively small number of artists working in just a few manufacturing centres in England, chiefly in Birmingham, the very centralised nature of the trade made for quite a high degree of standardization of decoration throughout England and Wales.

The corners, arches and centres of white dials were decorated according to fashion, with painted flowers, birds, fruit, stylised urns, geometrical patterns, shells, oriental features, sporting scenes (hunting, shooting), various 'foursomes' (seasons, continents, virtues, graces, apostles), ships at sea, picturesque ruins, castles, country cottages and rural scenes (with or without human figures), famous military personages or battles, and finally, in the second half of the nineteenth century, religious themes from the Old and the New Testaments.

Clockmakers could order their dials complete with standard painted decoration chosen from the manufacturer's list, like the following price list by Wilkes & Son of Whittall Street, Birmingham, issued in 1820:[2]

---

[1] *Grandfather Clocks*, pp. 105–8.

[2] Loomes, *White dial clocks*, p. 85.

Fig. 8. *Above:* Unrestored tidal dial by Thomas Jones of Newport *(Newport Museum and Art Gallery).* *Below:* Watercolour of the old bridge at Pontypridd, a favourite subject on local clock dials, from a watercolour by Thomas Hornor *(Glamorgan Archive Service).*

Additional Paintings and Movements charged as follows:	
Seasons, Quarters, Virtues, Elements, Etc	8s. 0d.
Landscapes or Figure Pieces in Arch	4s. 0d.
Single Figures painted in Arch	2s. 0d.
Adam and Eve to move	7s. 6d.
Ditto, if Serpent to move	10s. 0d.
Ship to move, Old Time, Etc	6s. 0d.
Swan's Neck to move	6s. 0d.
Boy and Girl swinging, Harlequin & Columbine, Etc	7s. 0d.
Shuttlecock and Battledore to move	7s. 0d.
Movements and Paintings of every Description	

The painted decorations on the dials are useful dating aids. For example, at the end of the eighteenth century the earliest painted dials were decorated with fruits, flowers or birds. Later on, the amount of painting on the dial increased, and decoration tended to fill the dial corners and the arch, and also much of the dial centre. Oriental features became fashionable in the first quarter of the nineteenth century, during the quite short period when Arabic numerals were preferred to Roman for the hours. Popular heroes, such as Nelson or Wellington, are easy to date, but some decorations remained popular for much of the nineteenth century, and so are of little use for dating.

These ever-popular decorations include the Four Seasons, usually portrayed by farm workers, some of them buxom women engaged on seasonal tasks, but sometimes by children e.g. bird-nesting, swimming, scrumping apples, and snowballing or skating. The Four Continents remained popular, again portrayed by female figures: Britannia with trident (Europe), a bare-breasted negress beside a lion-skin or elephant tusks (Africa), a Red Indian squaw complete with bow and arrows (America), and a voluptuous bejewelled ranee (India).

Because the dial decoration was usually chosen and ordered from a standard list, there was little scope for individualism, and recognisable local scenes are comparatively rare. When local scenes were requested, they were inevitably painted by people far away and unfamiliar with their subject and, understandably, the results are often inaccurate representations of reality. However, some South Wales clockmakers did request specifically local scenes, and perhaps supplied the artists with good illustrations to copy, for the dial decorations are good representations: Mumbles Head is a distinctive scene often observed on the dials of Swansea clocks; in east Glamorgan the favourite scene was the 'New Bridge' at Pontypridd; and in

Monmouthshire Tintern Abbey and Monmouth Bridge.

Pontypridd's 'New Bridge' over the Taf was built by William Edwards in 1754, after several previous failures. The bridge soon attracted great attention and in 1830 was described glowingly as 'this elegant structure ... exceedingly beautiful and picturesque ... a single arch, one hundred and forty feet in the chord, and thirty-five in height above the level of the river ...'.[1] More recently and prosaically it was described as 'one of the most dangerous and least serviceable of large bridges in Wales ... Its unique design, beauty and technical secrets have made it the most controversial bridge in Britain'.[2] It is no surprise that the Pontypridd bridge was a popular subject for dial decoration in the nineteenth century, the curved shape of the bridge lending itself admirably to portrayal in the arch or within the chapter ring of the dial.

Tidal dials are a special case. A few local moon-dial clocks also incorporate a tidal dial in the arch, calibrated to show the time of high water at a particular place, e.g. the Old Passage for the Severn ferry. A longcase clock at Newport Museum signed 'Thomas Jones, Newport' has a tidal dial showing 'High Water at Newport Bridge'. A clock by Henry Williams, Llancarfan, has 'High Water at Bristol Key'.

Signatures

As required by an Act of Parliament in 1698,[3] most clock dials bear a signature, that is the name of the maker or retailer, visible usually in the centre or on the periphery of the dial, or around the top of the arch, and generally accompanied by a place-name indicating the location of the clockmaker's workshop or shop premises.

In the period before mass production, a clockmaker either engraved his own name on the brass dial himself, or arranged for his name to be inscribed on the dial by an expert engraver. Although the quality of the engraving work varied considerably, there is usually no ambiguity about the reading or spelling of the maker's name or the

---

[1] *South Wales Illustrated* (1830).

[2] *Dictionary of Welsh Biography*, p. 198.

[3] Pryce and Davies, *Samuel Roberts*, p. 205.

place. Abbreviations are usually obvious: Edm^{d.} (Edmond or Edmund), and Grif^{h.} (Griffith). Occasionally, however, in the eighteenth century when engravers used cursive script instead of separate lettering, the eyes of a modern observer without some knowledge of palaeography may mistake certain combinations of letters. An example of this occurred when Iorwerth Peate simply misread the cursive *ew* in the name David Mathews as *en*, unfortunately took the surname to be Mathens, and perpetuated the error in the successive editions of his book.

Place-names on dials usually present little difficulty in identification, though some of the old forms and spellings of Welsh names may appear rather strange to modern eyes. *Llan* often appears as *Lan*, as in 'Lantrisant' or 'Lancarvan', 'Lawleston' is Laleston, 'Pont-Pwll' is Pontypool, and 'Monythusloyn' is Mynydd Islwyn. Curiosities include 'Lan-3-ssaint' and 'Lan-tris-Saint' for Llantrisant and 'P.P.' for Pontypool. It should also be remembered that for a while Newbridge was the usual name for the town that soon became and is now universally known as Pontypridd.

Exceptionally the name of the clock's owner may be inscribed on the dial instead of, or even as well as, that of the maker.

On painted dials, however, the signature often presents much greater problems. Mrs Tennant, an expert restorer of painted dials who works in north-east Wales, points out that the original signatures were usually put on by the dial manufacturer, not by the individual clockmaker.[1] This tended to result in lettering styles being fairly uniform in clocks assembled in different parts of the country at any given time, and in fairly generalised trends in lettering fashions over the course of the nineteenth century.

The chapter rings, numerals, signature and place-name were put on by pen and brush with water-based Indian ink. Over the years, many of the signatures on painted dials have been partly or completely erased by over-zealous cleaning, careless handling and rubbing, or even misguided wiping and washing. Thin lines and strokes, dots, and fine loops and scroll embellishments are the first things to be lost, especially from the earlier Gothic lettering. For example, a *t* may lose its fine crossbar and then resemble an *l*, or a capital *L* may lose its loops and tails and appear to be an *S*. If capitals such as *H* and *N* lose

---

[1] *Longcase Painted Dials: their history and restoration* (1995), pp. 57–8.

their cross strokes they may then be confused.

Though lost signatures can often be made out by a skilled and experienced eye under good oblique lighting or ultraviolet light, it is a sad fact that many initials and signatures on dials have been incorrectly restored by a combination of guesswork and ignorance. For example, a looped embellishment at the end of a name becomes interpreted as an *e*, giving a spurious form such as Kerne instead of the correct form Kern, or an initial *X* becomes incorrectly restored as $\mathcal{N}$ or $\mathcal{H}$.

Worse still is the situation where a name is quite deliberately inserted in place of a missing signature, in order to enhance the interest and hence the saleability of a clock. The worst case of all is unscrupulous fraud, where an existing genuine signature is quite deliberately erased and another more commercially attractive name is inserted in its place.

It should always be remembered, however, that even the genuine original signature of an individual clockmaker would usually, even inevitably, change over the period of his working life, as general lettering fashions changed. During the nineteenth century the fashion in clock signatures changed from ornate Gothic lettering with scrolls and loops in the early years to simpler capitals with serifs or upper and lower case letters, and finally, towards the end of the century, to plain poster-style capitals, upright or sloping, without serifs.

Rarely do dials show any information beyond the name and place of the clockmaker. William John of Newbridge (i.e. Pontypridd) was exceptional in indicating also his main occupation 'Cab$^{t.}$ Maker.' Occasionally, however, a clock dial may display some family history or social history. This happens especially in the case of presentation clocks, that is clocks presented on occasions such as coming of age, marriage or retirement, or as memorials in public buildings. An excellent example of this, though not a longcase, is a nineteenth-century round wall clock from a Rhondda chapel, seen recently for sale in Cardiff. This clock has a hinged dial and a fine fusee movement, rear-winding, and within its chapter ring of Roman numerals the dial bears the following inscription:

Presented by
John David Williams Esq. J.P.
of
Clydach Court, Trealaw
and
Mrs Mary Williams
White Hall
Trealaw
In Memory of
their
Father and Mother
John and Mary Williams
Cwmsaibron[1] Farm
Treherbert

---

[1] The modern version of the name is Cwm Saerbren.

Fig. 9. Three Cardiff clockmakers. *Above left* The Royal Arcade in the 1890s, with Leon Rosenberg's watch and jewellery shop (No 10) on the right; *Above right:* James Keir (1839–1921) outside his shop at 38 Castle Arcade, *c.* 1900. *Below:* Kaltenbach's shop at 23–24 Caroline Street, a family business from the 1860s to 1998, from a photograph of *c.* 1920. *(Stewart Williams).*

LOCAL CLOCKS ON PUBLIC DISPLAY

> The Clock Tower—appropriately enough—
> has a single iconographic theme: Time.
> J. Mordaunt Crook, *William Burges and the High Victorian Dream* (1981), p. 263

Cardiff Castle

It is fitting that the most impressive clock in South Wales should grace the Clock Tower of Cardiff Castle, in the heart of the capital city. Here the Marquess of Bute and the architect William Burges were clearly intent upon creating a visually outstanding and conceptually symbolic monument dedicated to time. Progress was rapid: the foundation stone of the Clock Tower was laid on 12 March 1869 and within two years the structure of the tower was completed. J. Mordaunt Crook, the biographer of Burges, describes it thus:

> The Clock Tower ... has a single iconographic theme. Time. Inside and outside, its decorative images portray the heavenly bodies and temporal divisions. The clock itself is flanked by statues—painted by Weekes and Smallfield—representing the seven planets, standing on pedestals carved with their respective signs of the zodiac. In the Winter Smoking Room the windows—the work of Weekes and Saunders (*c.* 1870)—represent the six days of the week, as enshrined in Saxon or Norse mythology: Moni, or the moon (in Scandinavian mythology Sol, the sun, is female and Moni, the moon, male); Tyr, a one-handed Northern Mars; Woden, or Odin, the terrible lord of Valhalla; Thor, the son of Odin, armed with a wondrous mallet and gauntlets of iron; Freyja, the Scandinavian Venus; and Satur, chief participant in the Ragnorök, or twilight of the gods. The seventh day is of course represented by the Sun, carved on the central boss of the room's vaulting. On the four walls and in the eight spandrels of the vault, are Weekes's paintings of the twelve signs of the zodiac. The four seasons appear in pictorial

form on the walls, as well as Lonsdale's murals of the origins of music and painting. Dawn, Sunrise, Sunset and Moonlight are depicted in the four corners. And the amusements of winter appear in sculptured form in the chimneypiece: Diana the Huntress, the Goddess of Heaven, being frequently identified with the Greek Artemis and Goddess of the Moon. In the Bachelor Bedroom above, we move from astrology to alchemy. The twelve signs of the zodiac appear in proxy form as the respective precious stones. The walls are labelled with thirty-two types of gem.[1]

The four-dial turret clock in the tower was made in 1870 by E. Dent & Co. of 61 Strand, London, a long-established firm celebrated for making Big Ben, the clock of the Palace of Westminster which had been installed in March 1859. Dent's firm acquired huge prestige as the maker of Big Ben, and was the obvious choice to make the Cardiff Castle clock for the Marquess of Bute. The bell in the tower at Cardiff was cast by John Warner & Sons in 1871, and bears the inscription 'E. Dent & Co, 61 Strand. Clockmakers to Her Majesty'.

At Cardiff Castle the Marquess of Bute had first-class clocks by the best London and Paris makers, and a fine bracket clock by John Thackwell of Cardiff, but these no longer adorn the rooms. The only clock of note still remaining on display is a large and anachronistic longcase clock by W.A. Allen of Cardiff with Gothic oak case, a long glass door to the trunk, and round brass dial with tumbling Arabic numerals. It is believed to be a Burges design, one of several made 'on approval' for the Marquess, *c.* 1890. An inventory of the Castle in 1931 shows that the servants' quarters then contained several humbler clocks bought from various Cardiff clockmakers during the nineteenth century, and including in the servants' hall an eight-day mahogany longcase clock with white enamelled dial by Ingram of Cardiff (then valued at £25).[2] This clock, by James Oliver Ingram, is now at Castell Coch, and its signature has apparently been incorrectly restored at some time, for it reads Ja$^{s.}$ D. Ingram.

---

[1] J.M. Crook, *William Burges and the High Victorian Dream* (1981), p. 263.

[2] Cardiff Castle Collection, Unnumbered MS.

LOCAL CLOCKS ON PUBLIC DISPLAY

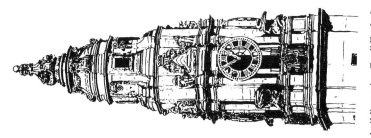

Fig. 10. Clocks on public buildings in Cardiff. *Left*: Cardiff Castle clock tower, containing a turret clock made by E. Dent & Co., London, in 1870 for the third Marquess of Bute *(Neil Turner Photolibrary)*; *Centre*: Former City Hall *(Cardiff County Council)*; *Right*: Pierhead Building, Docks *(Patricia Aithie)*.

Museum of Welsh Life, St Fagans

Local clocks are also on display to the public at the Museum of Welsh Life, where, in addition to the splendid clock of *c.* 1720 by Thomas Tompion in the long gallery of the Castle, several of the other buildings contain interesting clocks by local makers. In Kennixton farmhouse, an eight-day oak longcase clock of *c.* 1780, with brass silvered chapter ring and a ten-inch brass dial inscribed 'Richard Watkin' (of Merthyr Tydfil) stands beside the fireplace in the living room.

In Llwyn-yr-eos farmhouse there are a couple of mantle clocks; a round wall clock in the kitchen; a truly massive mahogany eight-day clock with a rolling moon dial in the arch, unsigned but late nineteenth century, in the living room (*y gegin orau*); and on the landing an oak eight-day longcase clock signed 'Thos Jenkins Dowlais', with a hunting scene in the arch, fruit in the spandrels, and a shell motif inlaid in the trunk door.

Also at St Fagans, in the re-erected terrace of houses from Merthyr Tydfil, the living room of the 1895 house contains an eight-day mahogany longcase clock signed 'James Thomas Merthyr' typical of the mid nineteenth century; and the 1925 house displays an ornate 'Vienna regulator', an accurate and attractive clock of the type that became popular in the middle of the last century and has remained so.

The extensive reserve collection of clocks and watches at St Fagans contains many items by local makers; these are not accessible to the general public, but can usually be inspected by researchers by appointment. The collection of watch-case papers includes examples by J.H. Beresford (Pontypridd), William Golding (Pontypool), B. Heitzman (Cardiff), Peter Heitzman (Pontypool), E. Kaiser (Cardiff), G. Rombach (Aberdare), and J.D. Williams (Merthyr Tydfil).

Cyfarthfa Castle Museum and Art Gallery, Merthyr Tydfil

Pride of place here must go to the regulator signed 'J.D. Williams. Merthyr Tydvil'; the round silvered main dial, thirteen inches in diameter, has small separate dials for the seconds and the hours. In its domed ebonised mahogany case, nearly seven feet high and two feet wide at the base, with mercury-filled pendulum, glass door and mirror back, this clock stood on display in jewellery shops in Merthyr Tydfil for over a hundred years, until it was finally bought for the museum in 1991.

The dining room of Cyfarthfa Castle contains a thirty-hour longcase clock signed 'Williams. Merthyr' (presumably William Williams), *c.* 1820–30. The twelve-inch dial has painted shell decorations in the arch and corners. The hood has swan-necked pediments, an eagle finial, and plain wood pillars with brass caps and bases. The case is light oak, with the 'Three Feathers' emblem neatly inlaid in the trunk door.

The 'Welsh Kitchen' display in the basement of the Castle includes a thirty-hour clock in a plain oak case, typical of Welsh country-style of the eighteenth century. The unsigned eleven-inch brass dial has a twelve-pointed star motif matted in the centre.

The other clocks on display at Cyfarthfa are an eight-day local longcase clock, signature unfortunately erased, with convex thirteen-inch arch dial with painted flower decorations; a factory time clock by the National Time Record Co. Ltd (St Mary Cray, Kent) for clocking workers in and out, from the Hoover factory at Merthyr; and a clockwork ballot box for voting in secrecy, invented by William Gould, which he exhibited at the National School Rooms, Merthyr, on 12 April 1870:

> The box is a most ingenious contrivance. It is twelve inches long, ten inches wide, and ten inches deep, has an aperture on the top for the insertion of the voting card, and an opening at the bottom for discharging its contents. At one end is a set of dials, which register the vote—one [dial] to register the single votes, the other [smaller dial] to record the hundreds, and these are worked by machinery set in motion by the insertion of the voting card.[1]

Joseph Parry's cottage at Merthyr Tydfil, where the musician and composer was born in 1841, displays a fine eight-day mahogany longcase clock of *c.* 1850, signed 'Powell & Davies, Merthyr Tydvil'. The thirteen-inch painted dial has a racing yacht in the arch, a sailing ship in each corner and, rather incongruously, two cottages inside the

---

[1] *Merthyr Express*, 16 April 1870.

chapter ring. The case is a typical example of the 'Bristol' style described above (p. 41), with inlaid stringing and parquetry bandings, and a notched domed hood with barley-twist (rope-twist) pillars with brass caps and bases.

Pontypridd Historical and Cultural Centre

The Centre, housed in the former Tabernacle Chapel, displays three clocks. One is a late nineteenth-century, eight-day longcase clock signed 'Evan Davies, Pont y pridd', better known as Myvyr Morganwg, bard and archdruid. It has a conventional painted thirteen-inch dial with buildings in the corners, and a lake view in the arch, but the case is unusual in having a carved crest to the dome of the hood, and also carved strips of wood inset into the canted corners of the trunk. This is the only known example of a Myvyr Morganwg clock, and was in the bard's family until 1995.

Another rather earlier eight-day longcase clock signed 'Wm. John, Cabt. Maker, Newbridge' (i.e. Pontypridd) is of *c.* 1840. The fine mahogany case has inlaid stringing and brass Corinthian cappings to the barley twist columns of the hood and trunk. The thirteen-inch dial has conventional flowers in the corners, and a country house scene in the arch.

Suspended from the chapel gallery is a round wall clock signed 'J. Crockett & Co, Pontypridd 1885'. It has a dial twelve inches in diameter, and a fusee movement.

Tredegar House, Newport

For years Tredegar House was home to many excellent clocks by famous London makers such as Samuel Town(e)son. Charles Octavius Swinnerton Morgan (1803–88), fourth son of Sir Charles Morgan and an authority on clocks and watches, bequeathed his own unique collection, including the celebrated Isaac Habrecht clock (1589), to the British Museum.

Today Tredegar House displays a few clocks of local interest. In the side hall is a thirty-hour longcase clock by Charles Vaughan of Pontypool, on loan from the Museum of Welsh Life. The square ten-inch dial has hour and minute hands and is signed 'C. Vaughan

Pontpool'. The plates of the movement have pieces cut out; the top pillars are conventionally round but the bottom pillars are rectangular in cross-section. The oak case is plain, with three acorn-type gilt finials on the front of the hood, and a caddy-style open-backed box behind them.

In the dining room is a fine eight-day mahogany longcase clock with thirteen-inch arched brass dial signed 'Henry Williams Lan Carvan'. The arch has a circular medallion engraved with an eagle and 'TEMPUS FUGIT'. This clock, of *c.* 1780, originally came from Tredegar House, but was sold in 1951, and is now on loan from the Museum of Welsh Life.

In a recess in the panelled corridor upstairs is another eight-day longcase clock by Henry Williams of Llancarfan. This clock was originally in the Tredegar House collection, and the case appears to have been specially made to fit the recess. The eleven-inch square brass dial is inscribed 'Henry Williams Lan carvan', with a seconds dial and date indicator. The oak case is 7ft 4in., with plain flat hood; the trunk door is 3ft 11in. long.

Newport Museum and Art Gallery

In the Town House is a longcase clock by Charles Vaughan of Pontypool, with square brass dial inscribed 'C. Vaughan Pontpool'. The clock has hour and minute hands, and the thirty-hour movement has Vaughan's usual brass-saving features (cut-away plates, and two iron pillars). The plain case is of plank, one inch thick.

The museum also displays a fine mantel clock with barometer and moon-phase dials, made for the office of Newport Harbour Commissioners in 1900.

Other clocks displayed include a mahogany longcase, *c.* 1840, with rounded dome top, and a convex painted dial signed 'Wm. Latch, Newport', and another early Charles Vaughan clock with brass dial signed 'C. Vaughan Pontpool'. This is a thirty-hour clock with a single (hour) hand, but in an ebonized oak case decorated by gilt painting and the date '1735' with a heart on the trunk door—perhaps originally given as a wedding present.

Fig. 11. *Left*: Evan Davies of Pontypridd, clockmaker, bard and archdruid, from the frontispiece of his book *Hynafiaeth Aruthrol* (1875). *Right*: J.D. Williams's shop, High Street, Merthyr Tydfil, c. 1905. *(Merthyr Tydfil County Borough Library).*

Torfaen Museum Trust, Pontypool

Two longcase clocks by Charles Vaughan of Pontypool are displayed. One is a thirty-hour clock with brass dial eleven inches square, signed 'Vaughan Pont pool', in a plain oak case, height 6ft 3in. It has hour and minute hands. The plates of the movement have pieces cut out; the top pillars are round and the bottom pillars rectangular in cross-section. The other clock is an eight-day longcase with a twelve-inch square brass dial, with hour and minute hands, a small seconds dial, and square date aperture. A plate on the dial is inscribed 'CHARLES VAUGHAN P.P.' The oak case is over 7 ft tall. All four pillars of the plated movement are brass, round in section.

SELECT BIBLIOGRAPHY

Brown, C. and Brown, M., *The Clockmakers of Llanrwst: pre-industrial clockmaking in a Welsh market town* (Wrexham, Bridge Books, 1993).

Fox, C.A.O., *An anthology of clocks and watches* (Swansea, The Author, 1947).

Greenlaw, Joanna, *Swansea Clocks: watch and clockmakers of Swansea and district* (Swansea, The Author, 1997).

Jarvis, P.H., 'Clockmaking in Glamorgan in the eighteenth and nineteenth centuries' (Unpublished Dissertation, Diploma in Continuing Education (Local History), University of Wales, Cardiff, 1995).

Jenkins, Gareth, *Monmouthshire Clockmakers* (Monmouth Archaeological Society, 1973).

Jones, G.M. and Scourfield, E., *Sully* (The Authors, 1986).

Parker, Paul, *Clockmaking in the Vale of Clwyd* (Mold, The Author, 1993).

Peate, Iorwerth C., *Clock and watch makers in Wales* (Cardiff, National Museum of Wales, 1945, 1960 and 1975).

Pryce, W.T.R. and Davies, T. A., *Samuel Roberts, Clock Maker: an eighteenth-century craftsman in a Welsh rural community* (Cardiff, National Museum of Wales, 1985).

Tennant, M. F., *Longcase painted dials: their history and restoration.* (London, NAG Press, 1995).

Waters, Ivor, *Chepstow clock and watchmakers* (Chepstow, The Author, 1972 and 1978).

Plate I. *Above:* Mahogany longcase clock by Powell & Davies, Merthyr Tydfil, *c.* 1850, with 'Bristol' features (i.e. characteristic of South Wales and the West of England (notched crest to hood, wavy border to door glass and rope-twist pillars); and detail of a mahogany case showing inlay and Corinthian capital. *Below:* Oak longcase clock by William Williams, Merthyr Tydfil, *c.* 1825, with 'Three feathers' emblem inlaid in trunk door.

Plate II. *Above left:* A London lantern clock of *c.* 1760. *Above right:* A modern German replica lantern clock (no South Wales example is known). *Below left:* Posted (birdcage) movement by Edmond Edmonds, Laleston, *c.* 1750. *Below right:* Brass dial of a single-hand clock by Edmond Edmonds, *c.* 1750, with chapter ring divided into quarter hours.

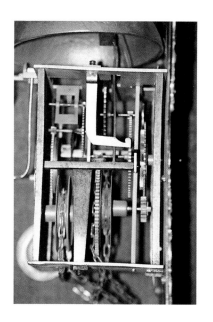

Plate III. *Above left:* Longcase clock by Charles Vaughan, Pontypool, *c.* 1750, in Newport Museum. *Above right:* Thirty-hour clock by Vaughan, *c.* 1760 *(Museum of Welsh Life)*. *Below left:* Regulator by J.D. Williams, Merthyr Tydfil. *Below right:* Mahogany longcase clock by Henry Williams, Llancarfan, *c.* 1780. Originally from Tredegar House and now displayed there on loan from the Museum of Welsh Life.

Plate IV. *Above left:* Mahogany longcase clock by James Oliver Ingram of Cardiff, *c.* 1860, at Castell Coch. *Above right:* Unsigned thirty-hour clock in oak case, on display at Cyfarthfa Museum and Art Gallery. *Below left:* Restored dial, signed Nicholas Shriber, Cardiff. *Below right:* American wall clock, signed F.J. Kaltenbach, Pontypridd, *c.* 1870.

Plate V. *Above left:* Clock case typical of 'Bristol' (i.e. South Wales and West of England) by William John, a cabinet-maker in Pontypridd, *c.* 1840. *Above right:* Mahogany longcase by Evan Davies (Myvyr Morganwg) of Pontypridd. *Below:* Two oak-cased clocks by Charles Vaughan of Pontypool; *left,* an eight-day clock at Torfaen Museum Trust; *right*, a thirty-hour clock at Tredegar Park, Newport.

Plate VI. Signatures of Edmond Edmonds (Laleston) and Powell & Davies (Merthyr Tydfil).

Plate VII. *Above:* Longcase clock by Richard Watkin, Merthyr Tydfil, in the living room of Kennixton Farmhouse at the Museum of Welsh Life *(Museum of Welsh Life)*. *Below:* Two of the four seasons (autumn and winter) used as decoration in the corners of a dial.

Plate VIII. Details of 'Four Continents' decoration used in the corners of a dial.

Comprehensive List of Clock and Watchmakers for Cardiff, the Vale and the Valleys

ALL THOSE LISTED HERE appear in directories, parish registers, gaol records or census enumerators' books up to 1900 as clockmakers, watchmakers or both. Some indication of status is also sometime given, e.g. apprentice, cleaner, jobber, or assistant. Where spans of years are given these are dates of birth and death where known, but in most cases indicate only the known period of horological activity.

The list contains all the names of local makers given by Iorwerth Peate (1975), amended and corrected, plus many new names found by further research, including those reported to the Museum of Welsh Life since 1975. The local census details are the only source for many of these people, who were not really 'makers' but merely employees of established clockmakers. Their names never appeared on the dial of any clock or watch. It should also be remembered that during the eighteenth century the spelling and form of names were not fixed. In particular, surnames such as Edmond, Evan, Griffith, Jenkin, John, Phillip, Walter, and William appeared with or without a final s. A selection of advertisements from local directories for some of those listed here has also been included in this section.

Peate's list of makers (1975) noted the locations and very brief details of many individual clocks and watches. This convention has not been followed here for three very good reasons: first, to preserve confidentiality; second, to avoid pointless repetition; and third, because clocks and watches, other than those in museum collections, are moveable items and therefore any location listing soon becomes hopelessly out of date and meaningless.

The names of those makers of signed clocks and watches actually seen by the writer, or reliably documented and known to exist, either in private ownership and/or public collections, are designated by an asterisk (*).

Allen, George James. Brynmawr and Dowlais 1887–95.
Allen, R. & Co. Ystrad Rhondda 1895.
* **Allen, W.A.** Cardiff. *c.* 1890.
* **Ansell, W.** Pontypool. *c.* 1870.
Anstee, Jonah. Bridgend 1867–87.
Anstee, Walter Charles. Laleston 1890s.
Arian, Israel. Treharris 1895.
Arnold, George. Merthyr Tydfil 1887.
Ashton, Thomas Heard. Bridgend 1871–5.
Attwell, William. Brynmawr 1875.

Bailey, Frederick. Pontypridd 1895.
Baker, John. Barry 1897.
Barbora, Charles. Cardiff. 19th century.
Barrett, Robert. Cardiff 1866.
* **Barry, James Trotter.** Cardiff 1841–75.
Bassett, Richard. Bonvilston.
Bassett, William. Bonvilston.
Bauman, Edward. Pontypridd 1871–87.
Bauman, Eugene. Pontypridd/Tonypandy 1875–87.
Beake, James William. Mardy, Rhondda 1895.
* **Beha, Leander.** Bridgend 1861–99.
Beresford, John Hewson. Pontypridd 1871–5. Later in Swansea.
Belcher, Jesse. Aberdare 1881.
Bellingham, Walter. Cardiff 1881.
Bernaschone, B. Merthyr Tydfil. First half of 19th century.
Berry, George. Cardiff 1858.
Best, George Christopher Henry. Cardiff 1881–7.
Best, Walter. Newport 1884.
Bevan, Rees. Porth 1895.
* **Beynon, Jane (?James).** Merthyr Tydfil 1868–87.
* **Beynon, Jeffrey.** Cardiff 1746–84.
Beynon, John. Llandaf/Cardiff 1770s.
Beynon, Lewis. Merthyr Tydfil 1841–61.
Bibby, John. Newport 1875.
Biddle, Edwin. Merthyr Tydfil 1871–87.

Biddle, Henry. Merthyr Tydfil 1861.
Biddle, Philip. Mountain Ash 1875–87.
Biggs, Bryant. Cardiff 1871–87.
Birkley, Primus. Newport 1868–99.
Blayward, John. Penarth 1871.
Bleibel, John. Cardiff 1875–99.
Bomash, T. S. Cardiff 1889.
Bond, William. Cardiff 1861. Apprentice.
Bonfiglio, Edward. Cardiff 1882.
Bonfiglio, Eugenio. Cardiff 1871. Watch-jobber.
Bowen, Owen. Merthyr Tydfil. *c.* 1795–1822.
Bowen, Thomas. Bridgend 1791–5.
Bradford, James. Cardiff 1865–75.
Bradford, Mrs S. Cardiff 1882.
Brauy, Charles. Cardiff 1871. Watchmaker's assistant.
Breeze, John Edward. Ynysybwl 1895.
Bright, Robert. Cardiff 1871.
Bright, Thomas. Cardiff 1871.
Brown, George. Cardiff 1871.
Brown, John. Llancarfan. 4 March 1778: apprentice to Henry Williams.
Brown, Nicholas E. Ogmore 1895.
Bunston, John Edwin. Pontypool 1884.
Burkle, Edward. Aberdare 1871.
Burns, Alfred. Dowlais 1887.
Buttery, John. Pontypool 1871.
Byrne, Daniel. Brynmawr 1868–75.

Campbell, Francis. Tonypandy 1881.
Canzi & Morris. Cardiff 1858.
Capella, J. Newport 1868–75.
* **Carey, Edward.** Pentre (Rhondda) 1887–90.
Carne, Robert. Pontycymer 1887.
Cary, Edward. Ystradyfodwg 1881. *See* Carey.
Clarke, James. Merthyr Tydfil 1861.
Clarke, John H. Cardiff 1881.
Clarke, William J. Mountain Ash 1881.
Clatworthy, Thomas. Ystrad Rhondda 1851.
Cletton, Bowen. Merthyr Tydfil 1841.

LIST OF MAKERS

Cohen, David. Merthyr Tydfil 1895.
Cohen, Lewis. Merthyr Tydfil 1895.
Cole, Charles. Cardiff 1879.
Cole, Frank. Cardiff 1881.
Cole, John. Newport. *c.* 1750.
Collier, D. Mountain Ash. Mid 19th century.
Collier (Collyer), O.N. Mountain Ash 1895.
* **Collings, H.** Cowbridge. 19th century.
Collings, Harris. Newport 1848–52.
Collings, Harry. Llantrisant 1887.
Collings, John. Cardiff 1855–85.
* **Collings, Joseph.** Cardiff 1854–75.
Collings, Joseph. Usk 1852.
Collings, Mrs Margaret Robertson. Cardiff 1882–7.
Collings, Samuel. Cardiff 1858–75.
Collins, George. Cardiff 1881.
Collins, Thomas W. Merthyr Tydfil 1881.
Colston, James. Cardiff 1887.
Connor, J.W. Treorchy 1875.
Connor, James. Mountain Ash 1871.
Cook, George. Merthyr Tydfil 1871.
Coombes, William James. Treorchy/Llwynypia 1892–5.
Coombs, W. Barry 1897.
Cooper, Richard. Mountain Ash 1881.
Corbett, Edward. Ystradyfodwg 1871.
Corbett, Edward. Cardiff 1881.
Cousens, Frederick. Cardiff 1871.
Crockett, Edward. Pontypridd 1887.
* **Crockett, John.** Pontypridd 1860–75.
Crouch, Henry Byron. Cardiff 1887–99.
Crouch, William Henry. Cardiff 1894.
Crowbat, Edward. Aberdare 1871.
Cuff(s), William Frederick. Ferndale 1886–99.

David, Evan. Coychurch 1841.
David, Thomas. Bridgend 1835–71.
Davies, Arthur. Ystradyfodwg 1881.
Davies, D.R. Treforest 1884.
Davies, David. Aberdare 1871. Apprentice in Merthyr 1861.
Davies, Evan. Aberdare 1864.

Also Dowlais.
Davies, Evan. Llanwonno 1861.
* **Davies, Evan.** Pontypridd (i.e. Newbridge) 1801–88 Bardic name Myvyr Morganwg; see p. 34.
Davies, Ivor A. Pontypridd 1851.
Davies, J.A. Ebbw Vale 1884. 'South Wales Watch Factory & Clock Supply: cheapest house in the trade.'
Davies, J.H. Cardiff 1858.
Davies, Job. Penydarren 1852–68. Dowlais 1871–5. Merthyr 1890. Aberdare 1895.
Davies, John. Llanwonno 1861.
Davies, John. Pontypool 1819–22.
Davies, John. Pontypridd 1871.
Davies, John L. Pontypridd 1895.
Davies, Michael. Pontypridd 1875.
Davies, Octavius. Glyntaff/Pontypridd 1861–81.
Davies, Owen Woolaston. Porth 1860–1936.
Davies, R.T. Merthyr Tydfil 1871.
Davies, Mrs Sarah. Dowlais 1887.
Davies, Samuel. Merthyr Tydfil 1871–5. Treherbert 1875. Aberdare 1879.
Davies, William. Aberdare 1881. Engraver.
Davies, William. Llanharan 1871.
Davies, William. Merthyr Tydfil 1825–65.
Davies, William Christmas. Mountain Ash 1887–99.
Day, James. Tredegar 1850–68.
Dold, Thomas. Merthyr Tydfil 1848–51.
Dorrer, Albert. Merthyr Tydfil 1871.
Dotter, Charles. Pentre (Rhondda) 1875–87.
Dotter, Joseph. Pontypool 1850.
Duffield, William. Ebbw Vale 1850–71.
Dufner, Hermann. Newport 1871–99.

* **Edmonds, Edmond.** Newton Nottage/Laleston. *c.* 1740. Died Aug. 1763.
Edmunds, Edmund. *See* Edmonds.
Edmunds, William. Cardiff 1841.
Elford, James. Cardiff 1874.

LIST OF MAKERS 69

* **Edwards, John.** Merthyr Tydfil.
Edwards, William. Treharris 1895.
Edwards, William Thomas. Ferndale 1886–7.
Eschle, Felix. Aberdare. *c.* 1840–88. Business continued by his widow Hannah Eschle 1895.
Evans, Caleb. Usk 18th century.
Evans, David. Llangynwyd. Early 19th century.
Evans, David. Newport 1852.
Evans, David. Pontypool 1849–84.
Evans, Ebenezer. Dowlais/Merthyr Tydfil 1868–81.
Evans, Edward. Merthyr Tydfil. 19th century.
Evans, George. Rhymney 1884.
Evans, John. Aberdare 1851–61.
Evans, John. Cardiff 1844.
Evans, John. Pontypool 1830–5.
Evans, John Cantor. Treherbert 1875.
Evans, Owen. Dowlais 1840.
Evans, R. & Co. Merthyr Tydfil 1884.
Evans, Rusky. Merthyr Tydfil 1887.
Evans, Thomas. Aberdare 1851–87.
Evans, Thomas. Pontypool 1849.
Evans, Thomas. Pontypridd 1895.
Evans, Thomas. Treharris 1895.
Evans, Thomas. Usk. First half of 18th century.
* **Evans, Walter.** Pontypool 1808–9.
Evans, William. Newport 1852.
* **Evans, William.** Pontypool. *c.* 1790–1820. Son of Caleb Evans. Apprentices: Thomas Mills and William Frost.

Faller, Frederick. Pontlotyn 1868. Rhymney 1871–5.
Farlong, William. Cardiff 1871.
Feale, William. Newport 1848–52.
Filippini, Frederick. Cardiff 1871 (apprentice). Pontypridd 1881–7.
Fisher, Alfred. Merthyr Tydfil 1881.
Flinn, John. Merthyr Tydfil 1881.
Flooks, Charles Henry. Merthyr Tydfil 1887–99.
Foley (Folie), Henry. Cardiff 1841. Merthyr Tydfil 1851–68.
Ford, Charles. Ystrad Rhondda 1895.

Ford, John. Ystradyfodwg 1881. Treherbert 1887.
Ford, L. Treherbert 1884.
* **Fox, Ralph.** Pontypool 1850–2.
Frampton, Charles George. Cardiff 1896.
Frank, Jacob. Pontypridd 1875.
Frantz, Broma. Aberdare 1861.
Fraser, Alexander. Cardiff 1897.
* **French, John.** Wenvoe. *c.* 1728–80.
* **French, Thomas.** Wenvoe. *c.* 1701–93.
* **French & David,** Wenvoe. Second half of 18th century.
Fresemeyer, Joseph. Brynmawr 1868–87.
Frewin, James. Abersychan 1849–71.
Frost, Edward. Newport 1830–52.
Frost, William. Newport 1822–30. apprentice to William Evans, Pontypool.
Fuhrer, Adolph. Treorchy 1895.
Fullwood, James. Newport 1868–87.
Furneaux, Thomas. Cardiff 1895.
Furtwangler, Albert. Cardiff 1881.
* **Furtwangler, Michael.** Cardiff 1855–71

Ganter, Felix. Cardiff 1851; Abersychan 1868–87.
Gardener, George. Cardiff 1891.
Gardner (Gardiner), Henry. Cardiff 1871–87.
Gefael, Lorenz. Merthyr Tydfil 1861.
Gelic, Deonysius. Cardiff 1881.
Gellrych, William. Cardiff 1881.
* **Godfrey, Moses.** Cowbridge 18th–19th centuries.
Goldberg, Ann. Cardiff 1884.
* **Golding, William.** Pontypool 1849–68.
Goldsworthy, William. Pontypridd 1861.
Goodman, David. Pontypridd 1844–68, Ystradyfodwg 1871.
Goodman, Samuel. Pontypridd 1871. apprentice.
Gottleib, Samuel. Cardiff 1891.
* **Gould, William.** Cefn-coed-y-Cymmer. *c.* 1810–70. Invented a

LIST OF MAKERS

clockwork ballot box.
Grant, Henry. Cardiff 1848–55.
Green, Josiah. Cardiff 1887.
Green, Woolf. Cardiff 1881.
Greener, F.J. Barry 1897.
Greening, James Henry. Penarth 1884.
Grieshaber, Anthony. Pontypridd 1881.
Grieshaber, Augustus. Pontypool 1868–99.
Grieshaber, John. Pontlotyn 1875.
Griffin, John. Cardiff 1867.
Griffith, Morgan. Bridgend 1861.
Griffiths, E. Tonypandy 1875.
Griffiths, G. J. Pontypool 1850.
Griffiths, George Geoffrey. Penarth 1887.
* **Griffiths, John.** Llantrisant. *c.* 1800.
Griffiths, John. Pentre (Rhondda) 1890.
Griffiths, Richard. Dowlais 1871.
Griffiths, William. Bridgend/Coity 1817–71.
Grishaber *see* Grieshaber.
Gruar, Thomas J. Newport 1875.
* **Guy, —** . Merthyr Tydfil 1824.
Gwinnett, Henry James. Cardiff 1887.

Halter, Albert. Cardiff 1882.
Hancock, John. Newport 1884.
Haniment, Edgar. Cardiff 1881. Apprentice.
Hannah, Henry Arthur. Cardiff 1892.
Harris, David. Merthyr Tydfil 1851.
* **Harris, Henry.** Mynyddislwyn. *c.* 1800.
Harris, Henry. Caerphilly 1840.
Harris, Lazarus. Merthyr Tydfil 1851.
Harris, Samuel. Cardiff 1869.
Harris, William. Cardiff. In 1734 this clockmaker was a prisoner in the County Goal.
Harris William. Leckwith, Cardiff 1861–66.
Harry, William. Michaelston-le-Pit 1850. Cardiff 1882.
* **Harvey, Hannah.** Pontypool 1835–42.

* **Harvey, John.** Pontypool 1822–30.
Hasenfratz, Antony. Cardiff 1861.
Hasenfratz, Charles. Cardiff 1866.
Haumphrey, Bernard. Energlyn 1871.
Havard, John C. Pontlotyn 1868.
Hayes, William Henry. Cardiff 1895.
Hayz, Nicholas. Merthyr Tydfil 1851.
* **Head, Joseph.** Cardiff. *c.* 1800.
Heitzman, Antony. Cardiff 1871.
Heitzman, Bernard. Cardiff 1851. Aberdare 1852.
* **Heitzman, Pius.** Cardiff 1851. Newport and Pontypool 1868–84. Penarth 1893.
* **Heitzman, Raymond.** Cardiff 1851–87.
Heitzman, Weibert. Aberdare 1881.
Heitzman, Wybert. Pontypridd 1895.
Henders, Thomas. Llantrisant 1861.
Hettich, Julius. Cardiff 1875–99.
Hill, Henry. Merthyr Tydfil 1881.
Hooper & Allen. Cardiff 1875.
Hopkins, John. Merthyr Tydfil 1881.
Hopkins, John Jenkin. Pontypridd 1895.
Hopkins, Rees. Aberdare 1856.
* **Hopkins, W.** Cowbridge. 19th century.
Horh, Adelbert. Merthyr Tydfil 1871.
Horowitz, Benjamin H. Penygraig 1895.
Howells, David. Ynysybwl 1895.
Hug, Constantine. Merthyr Tydfil, Troed-y-rhiw, Pontmorlais 1868–81.
Hug, Sebastian. Tredegar 1850.
Hughes, Robert. Merthyr Tydfil 1851.
* **Humel, Sigismund.** Cardiff 1851. Later in Swansea.
Humphrey, William David. Ferndale 1881.
Huntly, William. Cardiff 1853.
Hyman Brothers. Cardiff 1899.
Hyman, Jacob Harris. Dowlais 1895.

* **Ingram, James Oliver.** Cardiff. *c.*

1830–68.
Ingram, John. Cardiff 1837–48.
Ingram, John. Cardiff 1871–87. Penarth 1887–99.
Ingram, Robert. Cardiff 1845. Merthyr Tydfil 1861.
Isaacs, Abraham. Newport 1868.
Isaacs, E. Merthyr Tydfil 1873.
Ivey, Thomas. Merthyr Tydfil 1887.

Jackson, Alfred T. Pontmorlais 1868–73. Merthyr Tydfil 1875–99. Later in Swansea.
Jacobs, Hyman. Cardiff 1871–83.
Jacobs, Maurice. Cardiff 1881. Jobber.
James, John Lang. Cardiff 1890.
James, Thomas. Cardiff 1881.
James, Thomas. Abersychan 1884.
James, Thomas William, Tredegar 1868.
* **James, William.** Rhymney 1844. Merthyr Tydfil 1851. Cardiff 1882.
* **James, William.** Pontypool. 19th century.
Jenkin(s), Griffith. Llantrisant. Died 1729.
Jenkins, Evan Thomas. Ferndale 1895.
* **Jenkins, Thomas.** Dowlais 1835–52.
* **Jenkins, Thomas Jenkin.** Dowlais/Merthyr Tydfil 1844–87.
Jenkins, Thomas Rees. Merthyr Tydfil 1881. Assistant.
* **John, William.** Newbridge (i.e. Pontypridd) 1840–50. Cabinet-maker. Peate gives his name as Johns.
* **Johnson, Henry.** Cowbridge 1852–87.
Johnson, William Henry. Llantrisant 1871–87.
Jones, D.W. Merthyr Tydfil. Early 19th century.
Jones, David. Cowbridge 1830–5.
Jones, David. Aberdare 1861–8. Ystradyfodwg 1871.
Jones, David. Merthyr Tydfil 1760–1842.
Jones, David, jnr. Merthyr Tydfil 1835–52.
Jones, David. Pentre (Rhondda) 1875.
Jones, Evan. Gelligaer 1841.
Jones, Henry. Merthyr Tydfil 1851 (apprentice). Aberdare 1864.
Jones, Henry. Ynyshir 1883.
Jones, Henry. Cardiff 1893.
* **Jones, John.** Llandaf. 1645–1709.
Jones, John E. Cardiff 1881.
Jones, John Thomas. Ton Pentre 1895.
* **Jones, Llewelyn.** Dowlais. 19th century.
Jones, M. Newport 1884.
Jones, Matthew. Cardiff 1895.
* **Jones, R.** Gelli Deg.
Jones, Richard. Aberdare 1861–95.
Jones, Richard & Son. Pentre (Rhondda) 1884–95.
Jones, T. W. Tredegar 1852.
Jones, Thomas. Merthyr Tydfil. Before 1783.
* **Jones, Thomas.** Newport. c. 1800.
Jones, William. Cardiff 1881.
Jones, William. Merthyr Tydfil 1820–51.
* **Jones, William, jnr.** Merthyr Tydfil and Tredegar 1844–8.
Jones, William. Porth 1881.
* **Jones, William.** Tredegar. Died 4 Dec. 1841.
Jones, William Lloyd. Pontypool c. 1840.
Jordan, Thomas. Blaenavon 1868.
Jorona, Edward. Cardiff 1871.
Jorona, John. Cardiff 1871.
Joseph, Gwilym. Cardiff 1881. Apprentice.
* **Joseph, Thomas.** Merthyr Tydfil. Late 18th century.

Kaiser, Mrs Anne Maria. Cardiff 1884–7.
* **Kaiser, Elias.** Cardiff. c. 1865–87.
Kaiser, Mathias. Cardiff 1859–74.
Kaiser, Richard. Cardiff 1882. Penarth 1886.
Kaltenbach, Adrian. Cardiff 1868–95.
Kaltenbach, Edward. Cardiff 1868–87.
* **Kaltenbach, Francis Joseph.** Pontypridd 1868–87.
Kaltenbach, Frederick. Cardiff

1895.
Kaltenbach, George. Pontypridd 1871.
Karle, Adolph. Bridgend 1885.
Kearns, James. Merthyr Tydfil. Died 1840.
Keir, James. Cardiff 1839–1921.
Kerrick, George Townley. Penarth 1887.
Key, William. Cardiff 1807.
Kimich, Bernard. Newport 1852.
Kingetly see Kinstley.
Kinsey, David P. Newport 1868.
* **Kinstley, Joseph.** Tonypandy 1881. Pontypridd 1884–99.
Kinstley & Lehmann. Tonypandy 1875.
Knight, Benjamin. Cardiff 1855.
Knight, Thomas. Cardiff 1881.
Knight, William. Llantrisant 1881.
Korp, Michael B. Aberdare 1895.
Koos, George. Merthyr Tydfil 1871. Apprentice.
* **Koos, Leander.** Merthyr Tydfil 1868–99.
Kreiger & Co. Cardiff 1863.
Kuner, Isidor. Pontypridd/Porth 1868–87. Cowbridge 1874–5.

Langford, William. Cardiff 1881.
Latch, James. Newport. c. 1790.
* **Latch, William.** Newport 1830–71.
Lawley, Harry. Cardiff 1891.
Lea, Alfred William. Aberdare 1868–1924.
Leffler, Adolphe. Cardiff 1871. Assistant.
Lewis, Evan. Ferndale 1887–99.
Lewis, John. Tredegar 1850–1900.
Lewis, John Bateman. Merthyr Tydfil 1830.
* **Lewis, Rees.** Newbridge (i.e. Pontypridd) 1830–41.
Lewis, Thomas. Mountain Ash 1871.
Llewellyn, James. Merthyr Tydfil 1895.
Loeffler, Joseph. Pontypool 1884.
Long, James John. Cardiff 1887–92.
Long, Thomas. Penarth, 1871.
Long, William Henry. Cardiff 1867–75.
* **Loughor, John.** Kenfig/Bridgend. b. 1744, active 1765–90.

Luckett, Joseph. Cardiff 1891.
* **Lumley —.** Merthyr Tydfil. 19th century.
Lyons, Francis Henry. Cardiff 1877–99.
Lyons, Hubert John. Cardiff 1890.

Mabbett, H.G. Caerphilly. 1900.
Maison, Aristide Poterel. Brynmawr 1868.
Manley, William. Cardiff 1851–2.
Manly, William see Manley.
Manning, Francis Robert. Cardiff 1881 (apprentice). St Georges-super-Ely 1885.
Manns, R. Newport 1883–4.
Manorgoy, K. Cardiff 1871. Apprentice.
Marks, Jacob. Merthyr Tydfil 1875–81.
Marks, James. Treharris 1895.
* **Marks, Levi.** Bridgend. Before 1838.
Marks, Levi. Cardiff 1819–55.
* **Marks, Mark.** Cardiff 1829–52.
Marks, Michael. Cardiff 1813–22.
Marks, Nelson. Cardiff 1861.
Marks, Salomon. Cardiff. See Solomon Marks.
Marks, Samuel. Cardiff 1835.
* **Marks, Samuel.** Cowbridge 1820–44. Bridgend 1848–68.
Marks, Solomon. Cardiff 1822–82.
Marmont, William. Merthyr Tydfil 1861.
Masson, William. Cardiff 1881.
Maton, W.H. Cardiff 1891.
* **Matt, Matthew.** Cardiff 1861–71.
McKenzie, William. Cardiff 1881.
McMurray, John. Cardiff 1858.
Mendleson, P. Merthyr Tydfil 1895.
Meredith, Charles. Pontypridd 1881.
Meredith, William. Tredegar 1852.
* **Meredith, William.** Merthyr Tydfil 1852–99.
Meritz, Frank X. See Mertz.
Mertz, Franz Xavier. Treherbert 1875–87.
Miere, Lewis. Merthyr Tydfil 1841.
Miller, Philip. Tredegar 1868–71.
Mills, Henry. Merthyr Tydfil 1861.
Mills, Henry. Caerleon 1868–71. Newport 1884.

LIST OF MAKERS

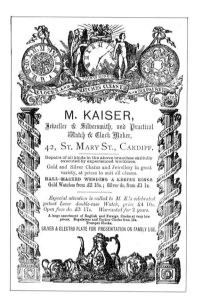

M. KAISER,
Jeweller & Silversmith, and Practical Watch & Clock Maker,
42, ST. MARY ST., CARDIFF.

Repairs of all kinds in the above branches skilfully executed by experienced workmen.
Gold and Silver Chains and Jewellery in great variety, at prices to suit all classes.
HALL-MARKED WEDDING & KEEPER RINGS.
Gold Watches from £2 15s.; Silver do. from £1 1s.

Especial attention is called to M. K.'s celebrated patent Lever double-case Watch, prices £4 10s. Open face do. £3 17s. Warranted for 2 years.
A large assortment of English and Foreign Clocks at very low prices. Regulateur and Cuckoo Clocks from 26s. Trumpet Clocks.
SILVER & ELECTRO PLATE FOR PRESENTATION OR FAMILY USE.

ELEGANT & USEFUL PRESENTS
AT
32, ST. MARY STREET.

R. HEITZMAN'S
CELEBRATED
Gold, Silver & Aluminium Watches,
All kinds of Clocks, and every description of Jewellery
AT REMARKABLY LOW PRICES.

GOLD WATCHES, IN 18 CARAT CASES.
3 guineas, 4 guineas, 5 guineas, to £40.
SILVER WATCHES.
21s., 30s., 42s., 3 guineas, to £10.
A LARGE STOCK TO SELECT FROM.

ELEGANT DRAWING-ROOM CLOCKS with Glass Shades complete.
25s., 42s. 63s., 4 guineas, to £20.
DINING-ROOM and HALL CLOCKS in Marble or Wood.
39s., 42s., 63s., to £10.
PARLOUR, BED-ROOM & KITCHEN CLOCKS.
3s. 6d., 10s. 6d., 15s. 6d., 21s., 30s., to £5.
CARRIAGE CLOCKS.
30s., 42s., 3 guineas, 4 guineas, to £12.
WARRANTED.

FASHIONABLE SOLID GOLD CHAINS AND ALBERTS.
25s., 42s. 63s., 4 guineas, to £20.
ALUMINIUM Do.
4s. 6d., 10s. 6d., 16s. 6d., to 20s. Assorted sizes and patterns.
A large Stock of 18 ct., 15 ct., and 9 ct. Gold Rings, Brooches, Ear-rings, and every description of Jewellery equally cheap. Quality guaranteed, and any article not approved of can be exchanged.
FREE AND SAFE BY POST.

OBSERVE!!
32, ST. MARY STREET,
(Opposite the Royal Hotel,) where the Clock projects from the House.

ESTABLISHED 1819.

Patronized by the Lords Commissioners of the Admiralty, and by the Honourable Board of Trinity, and the Royal Navies of England, France, Spain, and Holland.

SOLOMON MARKS,
CHRONOMETER, WATCH,
AND
Nautical Instrument Manufacturer,
JEWELLER, &c, &c.,
107, BUTE DOCKS,
CARDIFF.

Sole Agent for John Poole, Joseph Sewill, C. Frodsham, and French, of London.
Sole Agent for James Basnett, Richard Hornby, J. Sewill, Frodsham, Cairns, Campbell, and Jewitt of Liverpool.

CHRONOMETER MAKER TO THE ADMIRALTY.

New and Second-hand Chronometers constantly on Sale and Hire.

Chronometers Manufactured, Resprung, and Adjusted on the Premises.

CHARTS AND NAUTICAL STATIONERY.
IRON SHIPS SWUNG AND COMPASSES ADJUSTED.

T. J. WILLIAMS,

CHRONOMETER AND NAUTICAL
INSTRUMENT MAKER,
2, BUTE DOCKS,
CARDIFF.

Binacles, Compasses, Quadrants, Night Glasses, Aneriods, Sextants.

Steam Indicators and Gauges.

Admiralty & other Charts. Nautical Books and Stationery.

Adjuster of Iron Ship's Compasses.

Chronometers Let out on Hire.

LIST OF MAKERS

Mills, Thomas. Caerleon 1822; apprenticed to William Evans, Pontypool. Father of Thomas below.
Mills, Thomas. Caerleon. 1835–52, then Caerphilly (also Energlyn and Eglwysilan) 1852–81.
Mills, Thomas. Pontypridd 1852.
Minnett, George. Risca 1868.
Moore, Alfred. Cardiff 1884.
Moore, Arthur. Cardiff 1887–99.
Moretti, Frederick. Cardiff 1881.
Moretti, Rocco. Cardiff 1865–87.
Morgan, Anthony. Llangeinor. Buried 19 Sept 1767.
* **Morgan, George.** Cowbridge 1835–48.
Morgan, Howell. Blaengarw. *c.* 1890–1926.
Morgan, John. Merthyr Tydfil 1822.
Morgan, Joseph. Pontypridd 1871.
Morgan, Thomas. Newport. *c.* 1791.
Morgan, W. Pontypool 1850.
Morgan, William. Cardiff 1861.
Morley, William. Cardiff 1889.
Morris, James. Treherbert 1884.
Morris, P. Newport 1822.
Morris, William. Cardiff 1882–99.
Morselli, Geminiano, Cardiff 1865.
* **Munday, David.** Bridgend. Second half of 18th century.
* **Mundy, David.** Lawleston (i.e. Laleston). *See* Munday.

Nenn, James. Pontypridd 1841.
Newland, Charles. Newport 1873.
Newman, Albert Walter. Cardiff 1887. Barry 1897–8.
Newman, James. Aberdare 1871.
Newton, Frederick. Cardiff 1871.
Nolcini, John. Cardiff 1871. Clock-cleaner.
Nolcini, Joseph. Cardiff 1861–8.
Nolcini, P. Cardiff 1871. Clock-cleaner.
Novinski, Jacob. Penygraig 1895.

Owen, John. Merthyr Tydfil. Died 1838.
Owen, Owen. Merthyr Tydfil 1818. Shop robbed (*Cambrian* 18 Aug 1818).

Paff, M. Cardiff 1851. *See also* Pfaff.
Palmer, Charles. Newport 1852.
Palmer, George. Cardiff 1871. Apprentice.
Park, William. Aberdare 1861. Watch-finisher.
Park, William. Treorchy 1875.
Parry, David. Aberdare 1861.
Parry, Thomas. Cowbridge 1817.
Parry, W. Pontypool 1849–52.
Parry, W. M. Treforest 1861.
Parry, William. Aberdare 1848–9.
Parry, William. Newport 1868–81.
Parry, William. Pontypool 1835.
Pawsey, Harry. Cardiff 1895.
Pearce, Edward. Cardiff 1887–99.
Pedrazzini, B. Cardiff 1858.
Petrali, Angelo. Cardiff 1865–75.
Pfaff, Andrew. Merthyr Tydfil 1840–8.
Pfaff, Benjamin. Merthyr Tydfil 1861.
Pfaff, Romanus. Merthyr Tydfil 1848–68.
Phil(l)ip(s): note the variant spellings of this name; the Pentyrch and Llantrisant clockmakers were brothers.
Philip, —. Pentyrch 1772.
* **Phil(l)ip, David.** Llantrisant. Second half of 18th century.
* **Philips, John.** Pentyrch 18th century.
* **Phillips, —.** Pontypool.
Phillips, David and Isaac. Cardiff 1852.
Phillips, E. Cardiff 1855.
* **Phillips, John.** Llantrisant 18th century. *See* John Philips above.
Phillips, Joseph. Merthyr Tydfil. 19th century.
Phillips, Nathan. Merthyr Tydfil 1871.
Phillips, William. Tredegar 1839.
Piesold, W. Newport 1884.
Pitt, Thomas Serjeant. Newport 1842–52.
Plaskett, Charles. Cardiff 1863–4.
* **Polak, Samuel.** Pontypool 1825–50.
Poole, L. Cardiff 1881.
Porta, Peter. Cardiff 1868.
* **Powell & Davies.** Merthyr Tydfil.

c. 1810–40.
Price, David Lloyd. Beaufort 1844.
Price, Thomas. Pentre (Rhondda) 1887–95.
Price, William Morgan. Treorchy 1892.
Primavesi, F., and Son. Cardiff 1863–87.
Prin, Thomas. Cardiff 1867.
Pring, Charles. Cardiff 1881–3.
Pring, Jane. Cardiff 1881.
Probert, William. Aberdare 1852.
Purton, Joseph. Cardiff 1871. Penarth 1878.

Quelch, Richard. Ystradyfodwg 1881.

Read, John Mabyn. Pontypool 1849.
Reece, William Mortimer. Cardiff 1879.
Rees, George. Bridgend 1872.
Rees, James. Cardiff 1881.
Rees, John. Coychurch 1871.
* **Rees, Phillip.** Bassaleg. *c.* 1820–50.
Rees, Thomas. St Hilary 1842.
Rees, William. Cardiff 1835.
Reynolds, Morgan. Usk 1852–71.
Reynolds, Sebastian. Merthyr Tydfil 1841.
* **Richards, David.** Pontypridd 1851.
Richards, David. Pontllanfraith. Mid 19th century.
Richards, Frank. Cardiff 1881. Watch-jobber.
Richards, Llewellyn. Aberkenfig 1887.
Richards, Thomas. Cardiff 1789.
* **Richards, T.H.** Aberdare. 19th century.
Rieple, George. Bridgend. 19th century. Also at Swansea.
Riggs, Bryant. Cardiff 1881.
Ringwald, Benjamin. Merthyr Tydfil 1871. Cardiff 1881.
Ringwood, Benjamin. Caerphilly 1876.
Roberts, David. Bridgend 1895.
Roberts, Daniel Harris. Bridgend 1841–7.
Roberts, John. Cardiff 1858–68.
Roch, Benjamin. Pontypridd 1897.

Rogers, David. Glyntaff 1881.
Rogers, Lawrence. Llanhilleth 1776.
* **Rombach, George.** Aberdare 1871–81.
Rombach, Joseph. Cardiff 1851. Servant of R. Heitzman.
Rosenberg, Leon. Cardiff 1887–99.
Rosenbloom, Michael. Merthyr Tydfil 1851.
Rowland, Edward. Newport 1884.
Rowlands, Thomas. Ystradyfodwg 1884.
Rudda, James. Ystradyfodwg 1881.
Ruf, Markus. Pontypridd 1881.
Ruscombe & Co. Merthyr Tydfil 1884.

Salton, William. Usk 1884.
Samuel, Joseph. Cardiff 1852.
Samuel, William. Cardiff 1855.
* **Samuels, Barnet.** Merthyr Tydfil 1841.
Samuels, Moses. Cardiff 1858–75.
Saunders, Samuel. Bridgend 1870.
Scharlemaio, Augustus. Cardiff 1881.
Schiver, Charles. Cardiff 1851. Servant to R. Heitzman.
Schonart *see* Schonhardt.
Schonhardt, Andrew. Dowlais 1875–87.
Schonhardt, Augustus. Dowlais 1871.
* **Schonhardt, Frank.** Dowlais 1871–87.
Schreiber *see* Shriber.
Schubnell, Peter. Cardiff 1881.
Schwab, John. Aberdare 1871.
Schwalb, John George. Bridgend 1878.
Schwer, C. Cardiff 1871.
Schwer, Joseph. Brynmawr 1875.
Schwerer, Charles. Aberdare 1871–87.
Scott, Richard George. Llantwit Vardre 1881–5.
Seeley, Thomas Edmund. Mardy 1895.
Sherborne, J. Aberdare 1852.
Shill, H. Merthyr Tydfil 1836.
Shill, Samuel. Abersychan 1830.
* **Shill & Platnauer.** Pontypool. *c.* 1870.

LIST OF MAKERS

CHAS. NEWLAND,
Working Jeweller, Watch and Clock Maker,
ELECTRO-PLATER, GILDER, ETC.,
17, CHARLES STREET,
NEWPORT, MON.

W. H. WINSTONE,
Jewellery Manufacturer and
Working Goldsmith,
13 & 44, ROYAL ARCADE,
CARDIFF.
N.B.—Old Gold, Silver, Diamonds, Gold and Silver Lace, &c., Bought for Cash.
WATCH JOBBER TO THE TRADE.

The only Clerkenwell Watchmaker in Newport, is
R. MANNS,
7, LLANARTH STREET,
Watch Cleaned or New Mainspring, 2/6
Small Articles, Studs, Earrings, &c., Electro-plated while you wait from 3d. Spoons, Forks, &c., 6d. each.
Silver Watches from £1. Clocks from 5/6

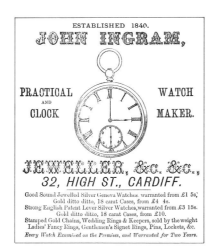

ESTABLISHED 1840.
JOHN INGRAM,
PRACTICAL AND CLOCK — WATCH MAKER.
JEWELLER, &c. &c.,
32, HIGH ST., CARDIFF.
Good Sound Jewelled Silver Geneva Watches, warranted from £1 5s.
Gold ditto ditto, 18 carat Cases, from £4 4s.
Strong English Patent Lever Silver Watches, warranted from £3 15s.
Gold ditto ditto, 18 carat Cases, from £10.
Stamped Gold Chains, Wedding Rings & Keepers, sold by the weight
Ladies' Fancy Rings, Gentlemen's Signet Rings, Pins, Lockets, &c.
Every Watch Examined on the Premises, and Warranted for Two Years.

LIST OF MAKERS

Shillito, Benjamin. Merthyr Tydfil. Jobber.
Shinn, Thomas. Mathern. Mid 18th century.
* **Shriber, Nicholas.** Cardiff 1761–40. Other spellings: Scriber, Scribor, Schreibner, Schreiber, Schriebner.
* **Siedle, Edward.** Merthyr Tydfil 1851.
* **Siedle, Edwin.** Merthyr Tydfil 1852.
Simmonds, Herbert. Aberdare 1865–68.
Skellorn, Herbert. Penarth 1895.
* **Skewis, James.** Aberdare.
Skinner, William. Aberdare 1887.
Smith, S. Taff's Well 1895.
Snook, William. Blaenavon 1884.
Solari, Giovanni. Cardiff 1887–99.
Southwood, James. Cardiff 1881. Apprentice.
Spears, William. Caerphilly 1895.
Spiridion, M. Newport 1883–4.
* **Spiridion, Wladyslaw.** Cardiff 1855–91.
* **Spiridion & Son,** Cardiff 1891.
* **Staurenghi, A.** Cardiff.
Steer, William. Newport 1884.
* **Stewart, Robert.** Newport 1848–56. apprenticed to William Latch.
Stilo, Henry. Cardiff 1871. Watchmaker's assistant.
Stradling, Lewis. Caerphilly 1791–5.
Strickland, George. Aberdare 1875.
Strong, Ebenezer George. Cardiff 1887.
* **Stroub, Constantine.** Treorchy 1875–87.
Sully, Henry W. Newport 1875–84.
Sully, John. Merthyr Tydfil 1861.
Sweetenburg, C. Merthyr Tydfil 1861. Assistant.

Tainsh Brothers. Cardiff 1882–8.
Tanner, Edward. Llancarfan 1771. Apprentice to Henry Williams.
Tayler, William. Newport 1884.
Taylor, Charles. Cardiff 1858–62.
Taylor, Henry. Cardiff 1871.
Taylor, Walter. Cardiff 1884.
Terry, Edward. Cardiff 1889.

Thackwell, Ann. Cardiff 1871.
Thackwell, Charles. Cardiff 1848–65.
Thackwell, Edward. Cardiff 1841.
* **Thackwell, John.** Cardiff 1740–1830.
Thackwell, Joseph. Cardiff 1841–71.
Thackwell, Thomas. Cardiff 1851.
Thatcher, Cornelius Octavius. Cardiff 1871–5.
Thatcher, Henry J. Cardiff 1868–93.
Thomas, Albert. Aberdare 1881.
Thomas, D. & Son. Ferndale 1884–7.
Thomas, D. E. Mountain Ash 1871. Jobber.
* **Thomas David.** Llantrisant 1780–1808.
Thomas, David E. Aberdare 1861–80.
* **Thomas, J.** Newport 1817. (*Hereford Journal* 15 Jan. 1817)
Thomas, James. Bridgend 1852–1917.
* **Thomas, James.** Merthyr Tydfil 1840–50.
Thomas, James. Tredegar 1850.
Thomas, James B. Cardiff 1861.
Thomas, John. Merthyr Tydfil 1861. Apprentice.
Thomas, John. Pontypridd 1871–87.
Thomas, John. Treherbert 1875.
Thomas, John Blount. Cardiff 1851 (apprentice), 1858.
Thomas, John J. Aberdare 1895.
Thomas, Rees. Bridgend, 1769 apprentice to David Munday.
Thomas, Samuel. Bridgend 1819–1900.
Thomas, Samuel. Aberdare 1895.
Thomas, Thomas. Merthyr Tydfil 1878.
Totter, Rowland. Pontypridd 1881.
Tratcher, Victor. Cardiff 1861.
Trenberth, Thomas. Cardiff 1881.
Trick, John. Merthyr Tydfil 1791–5.
Trick, Philip. Cardiff 1787.
Truscott, George. Coity 1841. Cardiff 1844. Newport 1868.
Tucker, R.G. Barry 1897.
Tucker, Robert. Cardiff 1855.

LIST OF MAKERS

Uphill, James Albert. Cardiff 1893.

* **Varrier (Verrier), Samuel.** Llanfair/Pentrebane. 1695–1765.
* **Vaughan, Charles.** Pontypool. *c.* 1730–95.

Vaughan, William. Newport 1852–84.

Ver(r)ier *see* Varrier.

Viner, Frederick Joseph. Cardiff 1895.

Vogt, Joseph. Aberdare 1861.

Vokes, Frederick. Newport 1875.

Voysey, John. Bridgend 1869–71.

Walker, James. Newport 1848–72.

Walter, Albert. Ystradyfodwg 1881. Treorchy 1887.

Walter, Harry. St Mellons. First half of 18th century. Son of William Evan Walter.

Walter, William Evan. St Mellons. 1679–1766.

Walters, George. Pontypool 1868.

Walters, Richard. Dowlais 1802.

* **Walton, Phillip.** Cowbridge. *c.* 1725–69.

Ward, Thomas Henry. Cardiff 1868–99.

Water(s), Harry. *See* Walter

Water(s), William Evan. *See* Walter.

* **Watkin, Richard.** Merthyr Tydfil 18th century.
* **Watkin, Thomas.** Merthyr Tydfil. Mid 18th century.
* **Watkins, —.** Merthyr Tydfil. 19th century.

Watkins, William. Blaina 1868–75.

Watson, Robert P. Cardiff 1882–7.

Watts, Benjamin. Cardiff 1887.

Way, Thomas. Newport 1852.

Webb, Walter. Merthyr Tydfil 1851.

Webber, Joseph A. Cardiff 1868–71.

Wehrley, Carl. Penarth 1887.

Weichert, William. Cardiff 1863–87.

Weighert *see* Weichert.

Weight, Edmund. Aberdare 1856–68. Merthyr Tydfil 1871–5.

Weight, Henry. Aberdare 1856.

Wells, Nugent. Newport 1868–99.

Wernet, August. Newport 1873–99.

Wernert *see* Wernet.

* **Whitehall, Edwin.** Newport 1848–75.

Whitehall, R. J. Newport 1884.

Wilkinson, J. Pontypool. *c.* 1800.

William, Charles 1841.

William, Walter. St Mellons. *See* Walter, William Evan.

Williams family. Llancarfan. The parish register shows that the following men named Williams were watch- or clockmakers in Llancarfan: Henry 1727–91; Edward 1705–63; another Edward born 1750; another Edward 1782–1860; Henry active 1820–30; and Thomas active 1830s.

Williams, C. and E. Newport 1883.

Williams, E.D. Cardiff 1871.

* **Williams, Edward.** Llancarfan and Caerphilly 1705–63.

Williams, Edward. Llancarfan 1782–1860.

Williams, Edward. Llandaff 1848.

Williams, Edward. Llantwit Major 1841.

Williams, Evan. Merthyr Tydfil 1822.

* **Williams, Evan.** Newport 1780–1830.
* **Williams, Evan Griffith.** Brynmawr 1844.
* **Williams, Griffith.** Newport 1760–95.
* **Williams, Henry.** Llancarfan. *c.* 1727–91. Apprentices: Edward Tanner 1771; John Brown 1778.
* **Williams, Henry.** Llancarfan 1820–30.
* **Williams, James David.** Merthyr Tydfil 1856–87.

Williams, John. Aberdare. Until 1875.

* **Williams, Thomas.** Llancarfan. First half of 19th century.

Williams, Thomas John. Cardiff 1851–93.

* **Williams, William.** Merthyr Tydfil 1822–35.

Williams, William. Ystrad Rhondda 1851.

Williams, William F. Merthyr Tydfil 1848.

Williams, William Ruffe. Newport

1848–52.
Willibald, John. Abertillery 1884.
Wilson, Alexander. Cardiff 1738–1824.
Wilson, William. Cardiff 1787–98. Son of Alexander Wilson.
Winstone, Harry. Cardiff 1887–99.
Winstone, William H. Cardiff 1873–83.
Winterhalder, L. Blaenavon 1884.
* **Winterhalder, Maximilian.** Pontypool 1868–84.

Woods, Oliver Ernest. Caerphilly 1887.
Wooton, George. Cardiff 1875.
Wright, Edmund. Merthyr Tydfil 1878. *See* Weight.
Wright, John. Merthyr Tydfil 1851.

Zahreneuger, Saloman. Pontypridd 1851.

APPENDIX

FACSIMILE OF THE CATALOGUE AND PRICE LIST ISSUED BY HOWELL MORGAN OF BLAENGARW, C. 1900

PRICE LIST

OF

Watches, Clocks, Jewellery,

ELECTRO-PLATE,

Challenge and Presentation Cups,

CUTLERY,

Musical Instruments, Bicycles, &c., &c.,

SOLD BY

H. MORGAN,

Watch and Clock Maker, Jeweller, &c.,

PARK HOUSE,

BLAENGARW.

Crick & Co., Printers, High Cross Works, Tottenham, London.

H. MORGAN, Watch and Clock Maker, Jeweller, Etc.

To Our Readers.

HAVING received many applications of late years for my Illustrated Price List of Watches, Clocks, Jewellery, etc., it is with much pleasure I beg to submit the present List to your kind consideration. It is, no doubt a competitive one, the prices are rare and cannot be beaten, and I feel confident that I can give you satisfaction in quality at moderate charges.

Having had many years personal experience in the trade, I can supply any of the following named goods on the shortest notice to any part of the United Kingdom with the prompest attention.

English, Swiss, and American Watches in Gold, Silver, or Metal.

All kinds of Jewellery in Gold, Silver-Plated and Gilt.

Sterling Silver Electro-plate and Britannia Metal Goods.

Presentation Marble Clocks and Bronzes, Brass, Wood and Nickel Clocks.

Weather Glasses in Wheel Barometers, Thermometers, Barometer Clocks, Aneroid Barometers, Opera, Field and Marine Glasses.

Fancy and Silver-Mounted Leather Goods.

Table and Pocket Cutlery.

Musical Boxes, Musical Albums, Violins, Violas, Banjoes, Accordians, Concertinas, and every kind of Musical Instruments.

Park House, Blaengarw, South Wales.

H. MORGAN, Watch and Clock Maker, Jeweller, Etc.

Harmoniums, American Organs, and Pianos, a special printed Warranty with every one. All Pianos warranted for 10 years by the manufacturers.

Sewing Machines of every description and size, with 5 years Warranty.

Cycle Machines and Tricycles of the finest workmanship.

We are in a position to execute every description of Watch, Clock and Jewellery Repairs, Mounting and Electro-plating, expeditiously and at the cheapest possible rates.

Being a native of South Wales, I can supply any article or instrument mentioned in this Price List cheaper than any firm in England.

Remittance may be sent by P.O.O.'s, Postal Orders, Cheque or Cash, in Registered Envelopes.

Packing Cases must be returned within 7 days carriage paid, or will be charged for, unless arrangements are made for large orders, then cases are free, and sent per goods train at a special low rate. Small packages not exceeding 11 pounds sent per Parcels Post Exceeding 11 pounds per passenger train.

It is impossible to give more than an idea of the extensive variety I keep in stock.

It is advisable when making purchases for customers to call personally, if possible ; this course will enable them to see and examine for themselves the articles illustrated.

If not convenient to call please quote number and design in List with all orders, and will answer per return giving reason for the enormous difference in price, quality and design.

Park House, Blaengarw, South Wales.

H. MORGAN, Watch and Clock Maker, Jeweller, Etc.

Comparison of price alone is most delusive.

Customers will find this List of some service when unable to call. An Order for the cheapest article or low-priced Watch in the List will receive the same attention as the highest class of goods.

Trusting to be favoured with your orders, which shall have my personal care and attention,

>With Compliments,
>
>I remain, yours truly,
>
>*H. Morgan.*

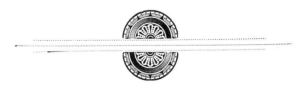

Park House, Blaengarw, South Wales.

H. MORGAN, Watch and Clock Maker, Jeweller, Etc.

Gold Watches in reach of all!

Called Gold Filled Cases, and are made of what is known as Rolled Plate Gold.

A Thick Plate of 10ct. Gold outside, and a thinner plate of 10ct. Gold inside. Composition plate in the middle, so you can see the cases are very strong having three plates together.

10-Carat Cases, Price £5 0 0

For higher prices the same Watch can be obtained with 14-ct. Plates and Composition Marked Gold Filled Cases.

14-Carat Cases, Price £6 10 0

There is a Printed Certificate with each Case in 14-Carat to wear for 20 Years.

I have no doubt these Watches will last (with fair wear) for 30 years although the certificate only mentions 20. If in 30 years the Gold gets thin, the case is then strong through the middle plate.

14-Carat Cases, Price £6 0 0

My reason in calling the customer's attention to this Watch, is because it is equal in appearance and movement to a watch costing thrice the money. So take advantage of this offer and buy a Watch for little money.

14-Carat Cases Price £6 0 0

Every Watch Warranted free of charge for Five Years if fairly used.

Park House, Blaengarw, South Wales.

H. MORGAN, Watch and Clock Maker, Jeweller, Etc.

Gentlemen's Silver Watches.

1

2

3

No. 1.—GENT.'S WATCH in Real Silver Cases, Hall-marked, with Lever Movement and Compensation Balance, 42/-, 48/-, 55/- and 60/-

No. 2.—GENT.'S SILVER WATCH, with fine quality Movement, full Jewelled, 30/- 35/- 40/- 45/-

No. 3.—GENT.'S HORIZONTAL WATCH, in Silver Cases, Hall-marked, ¾-plate Movement with flat Crystal Glass. Warranty for Three Years, price £1.

EACH WATCH WARRANTED FOR THREE YEARS.

Park House, Blaengarw, South Wales.

H. MORGAN, Watch and Clock Maker, Jeweller, Etc.

Gentlemen's Silver Watches.

4

5

No. 4—GENT.'S ¾-PLATE KEYLESS, 3 Pairs Extra Jewelled, in sound English Hall-marked Cases, with Five Years Guarantee £4 5s.

No. 5.—GENT.'S KEYLESS WATCH, English Lever Escapement, Demi-Hunter, Jewelled in 6 holes, Five Years Guarantee, 53/-, 60/-, 65/-, 70/-, 75/-

No. 6.—GENT.'S LEVER, this is one of the best and cheapest Watches for the price, in English Hall-marked Cases, with Steel Balance, 55/-

Warranted English Make throughout.

Ditto.   With Compensation Balance, and thicker Cases 63/-

6

Park House, Blaengarw, South Wales.

H. MORGAN, Watch and Clock Maker, Jeweller, Etc.

Ladies' Silver Watches.

7 8 9

No. 7.—LADIES' KEYLESS HORIZONTAL WATCH, in Sterling Silver, Fancy dial and Engraved Cases 18/-

Ditto.   Fully Jewelled and well finished, 25/-

No. 8.—LADIES' HORIZONTAL WATCH, in Stout Silver Cases, 28/- 30/-, and 35/-

Ditto.   Best finish, deeply engraved, fully Jewelled, 50/-

No. 9.—LADIES' KEYLESS DEMI-HUNTER, with Sterling Silver Cases beautifully engraved, ¾-Plate, 45/-, 50/-, 55/-, 60/-

ALL WATCHES WARRANTED FOR THREE YEARS.

Park House, Blaengarw, South Wales.

H. MORGAN, *Watch and Clock Maker, Jeweller, Etc.*

Metal Case Watches, Extra Strong.

10

11

No. 10.—H. MORGAN'S WORKMAN'S WATCH, fitted only in Keyless, this Watch is made expressly for underground use and rough wear. Lever Movement and timed ready for use. There is no need to open this Watch, except for cleaning or repairing, the Watch is guaranteed for two years, price 16/6 in Nickel Case.

Ditto. Same Watch in Real Silver Cases, 35/-

Ditto. Extra Stout Cases, Compensated Balance, 40/-

No. 11.—GENT.'S CENTRE SECONDS STOP WATCH, Metal Case, 14/- to 18/-

Ditto. With Lever Movement, good finish, fully Jewelled, £2 15s., £3, and £3/10.

THESE WATCHES ARE WARRANTED FOR TWO YEARS.

Park House, Blaengarw, South Wales.

H. MORGAN, Watch and Clock Maker, Jeweller, Etc.

Gentlemen's Silver Alberts.

All Engravings of Alberts show the exact size, that will be supplied at the price, but short pieces only are shown.

Each full length Chain is fitted with Patent Swivel, Bar and drop piece.

Every Chain is guaranteed to be the quality represented.

Chains (except Fancy Patterns) are not stamped, but the quality is quite as good as those that are stamped.

If any customer should require a smaller and cheaper Chain than what is represented, he may choose it from this List, as prices are for the lightest as well as the heaviest makes of each pattern. A separate price is given to each Albert.

We have a large quantity of designs in stock, which would be impossible for us to show in this List.

When enclosing an order for any article, customers should carefully quote number and price, and if the exact size as shown by engraving is required, prices at end of each Albert, so as to avoid mistakes.

Either of the following patterns may be had in Silver or Gold:—

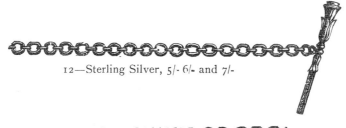
12—Sterling Silver, 5/- 6/- and 7/-

13—Sterling Silver, 14/- and 16/-

14—Hall-marked Silver, 5/6

Park House, Blaengarw, South Wales.

H. MORGAN, Watch and Clock Maker, Jeweller, Etc.

Gentlemen's Silver Alberts.

15   14/-

Hall–marked 7/6 to 15/-

16   6/6, 7/6, 8/6, Fancy.

Sterling Silver from 10/-, 15/-, 25'-

17   8/6

Sterling Silver, 8/6, 12/6, 15/-, 18/-

18   11/-

Hall–marked Silver Curb Links, from 6/- up to 25/- each.

19   9/9

Hall–marked Silver, from 8/- to 30/-

20   16/-

Hall–marked Silver from 6/6 to 28/-

21   10/9

Hall–marked Open Curb, from 7/- to 20/-

22   18/-

Hall–marked Silver, 16/- to 30/-

Park House, Blaengarw, South Wales.

H. MORGAN, Watch and Clock Maker, Jeweller, Etc.

Real Silver Brooches.

NEW DESIGNED GOODS.

Please note that each price represents the same pattern, but the higher price is for a heavier make.

Any article, however small will be forwarded to any part of the United Kingdom, safely packed and registered.

I stock an endless variety of Brooches which it would be impossible to show in this List, so I have put in what I think the most becoming designs between the large and extremely small Brooches, which will never be out of fashion. So that any order that you may have for patterns not in this list, forward on to me and I shall be pleased to execute the same.

I have a choice selection always on view.

23—Spray Brooch, Oak and Ivy,
2/6 and 3/6.

24.—Hall Marked Silver,
2/6 and 3/-

25—Rich Brilliant,
3/9 and 4/9.

26.—Paste Star Silver Brooch, Hall-marked.
4/6 and 5/9.

27.—Silver and Paste,
4/- and 5/6.

28.—Silver, Hall-marked,
3/6 and 4 6.

29.—Silver Swallow Brooch, Hall-marked,

30.—Silver Hall-marked.
3/- and 4/6.

Park House, Blaengarw, South Wales.

H. MORGAN, Watch and Clock Maker, Jeweller, Etc.

Real Silver Brooches.

31.—Brilliant Star,
7/6.

32.—Good Luck, Real Silver
2/4 and 3/6.

33.—6/9.

34.—5/6.

35.—Hall-marked Silver,
4/9 and 5/9.

36.—Crown Brooch, in
H.M. Silver 2/9 and 3/9

37.—Hall-marked Silver,
3/6 and 4/6.

38.—3/6.

39.—4/6.

40.—3/-

41.—3/6.

42.—4/-

43.—3/-

The Fashionable Curb Brooches I have in stock at same prices in Hall-marked Silver.

Park House, Blaengarw, South Wales.

H. MORGAN, Watch and Clock Maker, Jeweller, Etc.

Gent.'s Silver Pendants, &c., for Alberts.

The following illustrations are of the finest quality and workmanship, and at prices which I think will command an increasing sale.

44.—Hall-marked Silver, 2/6

45—6/-

46—Hall-marked Silver 5/-

47.—Hall-marked Silver, 6 9.

48.—Hall-marked Silver 6/6

49—Hall-marked Silver, 6/6

50.—2/-

51.—1/6

52.—Masonic Jewel. 3/-

53—2/9

Park House, Blaengarw, South Wales.

H. MORGAN, *Watch and Clock Maker, Jeweller, Etc.*

Gent.'s Silver Pendants, &c., for Alberts,

SUITABLE FOR STRONG AND HEAVY ALBERTS.

A smaller size than those illustrated, I can supply much cheaper.

55—Hall-marked Silver 4/6

56—9/6.

57—Hall-marked Silver, 6/6

58.—Real Silver, Hall-marked, 5/9

59.—5/6.

SUITABLE FOR LADIES.—I have a large assortment of Silver Pendants and Charms both in large and small sizes, on view, and on application, will forward samples of the same on approbation.

Park House, Blaengarw, South Wales.

H. MORGAN, Watch and Clock Maker, Jeweller, Etc.

Gent.'s Snake and Signet Rings,

MARKED AT POPULAR PRICES.

These Snake and Signet Rings are the latest frshion and massive. Lighter Rings of the same pattern from 7/6

60.—Hall-marked Signet, 9 ct. 18/- 18 ct. 35/-

61.—Snake Ring, Hall-marked, 9 ct. 24/- 18 ct. 42/-

62.—Buckle Ring, 9 ct. 20/· 18 ct. 35/-

63.—Hall-marked Signet. 9 ct. 30/- 18 ct. 52/-

64.—Hall-marked Twist Ring. 9 ct. 18/6 18 ct. 38/-

65.—Gent.'s Signet, 9 ct. 30/- 18 ct. 42/-

66.—Massive and broad Signet, 9 ct. 27/- 18 ct. 42/-

67.—Hall-marked Twist Ring. 9 ct. 23/- 18 ct. 40/-

68.—A grand Signet, 9 ct. 21/- 18 ct. 38/-

69.—Broad Signet, Hall-marked, 9-ct- 35/- 18 ct. 60/-

70.—Hall-marked Snake 9 ct. 15/- 18 ct. 33/-

71.—Large Signet, 9 ct. 25/- 18 ct 39/-

Park House, Blaengarw, South Wales.

H. MORGAN, Watch and Clock Maker, Jeweller, Etc.

Ladies' Fancy Rings and Keepers.

72.—9 ct. 12/- 22 ct. 15/- 17/- 20/- 32/- 40/- 50/- and 60/-
73.—9 ct. 18/; 18 ct. 28/-
74.—9 ct. 12/- 18 ct. 26/-
75.—9 ct. 16/- 18 ct. 30/-

76.—9 ct. 12/- 18 ct. 24/-
77.—9 ct. 12/- 18 ct. 24/- half tablet Keeper.
78.—9 ct. 20/- 18 ct. 40/-
79.—9 ct. 18/- 18 ct. 20/-

80.—9 ct. 9/- 18 ct. 18/-
81.—18 ct Diamond 45/-
82.—9 ct. 18/- 18 ct. 36/-
83.—9 ct. 15/- 18 ct. 30/-

84.—15 & 18ct. Diamond, 3 diamonds, 60/- to 140/-
85.—9 ct. 10/- 18 ct. 20/-
86.— 9 ct. 11/- 18 ct. 22/-
87.—9 ct. 13/- 18 ct, 26/-

88.—9 ct. 14/- 18 ct. 28/-
89.—9 ct. 10/- 18 ct. 20/-
90.—9 ct. 15/- 18 ct. 30/-
91.—9 ct. 16/- 18 ct. 36/-

92.—9 ct. 14/- 18 ct, 28/-
93.—9 ct. 8/6, 18 ct. 17/-
94.—9 ct. 16/- 18 ct. 32/-
95.—9 ct. 12/- 18 ct. 24/-

96.—9 ct. 17/- 18 ct. 34/-
97.—9 ct. 18/- 18 ct. 40/-
98.—9 ct. 13/- 18 ct. 26/-
99.—9 ct. 12/6 18 ct. 25/-

100.—9 ct. 14/- 18 ct, 28/-
101.—9 ct. 18/- 18 ct. 36/-
102.—9 ct. Buckle, 15/- 18 ct. 30/-
103.—9 ct 10/- 18 ct. 19/-

104.—9 ct. 14/- 18ct. 28/-
105.—9 ct. 17/- 18 ct. 38/-
106.—9 ct. 14/- 18 ct. 28/-

Park House, Blaengarw, South Wales.

H. MORGAN, Watch and Clock Maker, Jeweller, Etc.

Gentlemen's Gold Watches.

107

108

No. 107.—GENT.'S 14 ct. GOLD PATENT LEVER Escapement, Keyless Watch, Crystal Glass, Compensation Balance, engine turned Cases, £6.

Ditto: In 18 ct. Cases 3 years guarantee, £7 10s.

No. 108.—A Key-Winding ENGLISH LEVER Jewelled, in Stout Cases, Crystal Glass, Compensated Balance fitted in Crystal Glass Bezels, 3 years guarantee, £6 15s.

No. 109.—GENT.'S ENGLISH PATENT LEVER, Jewelled, with Compensated Balance, thoroughly good strong movement, in 18-ct. Gold Cases, with 5 years guarantee, £12. With Gold Dial £13 10s

109

Park House, Blaengarw, South Wales.

H. MORGAN, Watch and Clock Maker, Jeweller, Etc.

Gentlemen's Gold Watches.

CHRONOMETER BALANCE IN STRONG CASES.

110

111

112

No. 110.—GENT.'S GOLD WATCH, Non-Keyless, in 18 ct. cases, English Patent Lever, medium size very attractive appearance, Capped and Jewelled, 7 years Warranty, £16 16s.

Ditto. 18 ct. fitted with Gold Dial. £18, £20, £22, £24.

No. 111.—GENT.'S 18 ct. ENGLISH PATENT LEVER, Keyless, extra fine movement, in Hunting Cases, engine turned or plain, white dial, with heavy solid cases, 7 years Guarantee. £26.

No. 112.—GENT.'S KEYLESS ENGLISH LEVER, Demi-Hunting Cases in 18 ct., highly recommended, a thoroughly good serviceable Watch, with escapement on end stones, £25 10s.

Park House, Blaengarw, South Wales.

H. MORGAN, Watch and Clock Maker, Jeweller, Etc.

Ladies' Gold Watches.--A Speciality.

All Watches are Guaranteed to be the best quality.   Printed Warranty from 3 to 5 years.

112A   113   114

No. 112A.—LADIES' HORIZONTAL WATCH, Crystal Glass, 14 ct. Gold 50/-
No. 113.—LADIES' KEYLESS, Gold, Superior Finish Movement, Crystal Glass 63/-
No. 114—LADIES' 14 Ct. Stout Cases, Fully Jewelled, Crystal Glass, 84/-

115   116   117   117A

No. 115.—LADIES' 18 Ct. GOLD KEYLESS, extra good finish, fully Jewelled 95/-
No. 116.—   ,,   ,,   with Patent Lever Escapement, Key-winding, £5 5s.
No. 117.—LADIES' DEMI-HUNTER, Keyless, Patent Lever, best finish, £7.
   No. 117A.—LADIES' ENGLISH PATENT LEVER Gold, Capped and Jewelled, Cases deeply engraved, best finish, Presentation Watch, £11, £13, £15.

Park House, Blaengarw, South Wales.

H. MORGAN, Watch and Clock Maker, Jeweller, Etc.

English Patent Levers for Gents.

118

119

120

No. 118.—H. MORGAN'S Celebrated ENGLISH Levers, Sterling Silver Cases, double bottom with improved index for hairspring, fitted with light steel balance, Capped and Jewelled, polished pivots, Warranted for 5 years, £3 5s.

Ditto. Better Finish with Gold Balance, £3 10s.

No. 119.—GENT.'S WALTHAM, Patent Pinion, Compensated Balance, good strong Lever, 5 years Warranty £3.

These Watches have my name and address on dial.

No. 120.—GENT.'S ENGLISH LEVER, medium size, Sterling Silver Cases, Compensated Balance, Capped and Jewelled, 5 years Warranty, £4 4s.

Ditto with Silver Dial and Gold figures, £4 12s.

Park House, Blaengarw, South Wales.

H. MORGAN, Watch and Clock Maker, Jeweller, Etc.

English Patent Levers for Gents.

121

122

123

No. 121.—GENT.'S LEVER, English Patent, Chronometer Balance, well finished, air, dust and damp tight, Fuze Movement, extra Jewelled, best Polished Pivots, Enamelled Dials, (with my name on dial,) in strong Silver Cases, £5 10s., £6 10s., £7.

No. 122.—GENT.'S KEYLESS DEMI-HTUNER ¾-Plate English Lever, extra Jewelled, Plain or Engine-turned Case, £5 10s., £5 15s., to £7.

No. 123.—GENT.'S PATENT LEVER. medium size, Fuze Movement, Capped and Jewelled, with Gold Balance, £4 15s.

Ditto. Fitted to Chronometer Balance, with Silver Dial, £5 5s.

Park House, Blaengarw, South Wales.

H. MORGAN, Watch and Clock Maker, Jeweller, Etc.

Gentlemen's Gold Alberts.

New and Fancy deeigns in 9, 15 and 18 ct. Gold.

The high-class character of these Chains are as well as other goods supplied by me, that they are at once discerned. The prices quoted show the immence saving effected by dealing with me. They could not be purchased anywhere else at the prices.

All Engravings of Alberts show the exact size, that will be supplied at the price, but short pieces only are shown.

Each full length Chain is fitted with Patent Swivel, Bar and drop piece.

Every Chain is guaranteed to be the quality represented.

Most of these Patterns are Hall-marked on every link, but Fancy Patterns the Govern- Assay Office will not stamp, therefore are marked on one end only, and the quality of each guaranteed.

If any customer should require a smaller and cheaper Chain than what is represented, he may choose it from this List, as prices are for the lightest as well as the heaviest makes of each pattern. A separate price is given to each Albert.

We have a large quantity of designs in stock, which would be impossible for us to show in this List.

124—9 ct. £3, £3 10s.

125—9 ct. £4 10s.

126—9 ct. £4.   Hall-marked £4 10s.

Park House, Blaengarw, South Wales.

H. MORGAN, Watch and Clock Maker, Jeweller, Etc.

Gentlemen's Gold Alberts.

127.—Hall-marked in 9 ct. £3 3s., £3 15s., £4 10s., 15 ct. £5 5s., £5 18s., £7 8s. 6d.

128.—9 ct. £2 10s., £3, £3 10s.   15 ct. £4 3s., £5, £5 7s.

129.—9 ct. £3, £3 10s., £4.   15 ct. £5, £5 17s. £6 13s.

130.—Hall-marked 9 ct. £3 10s., £4, £4 10s.   15 ct. £5 17s, £6 13s., £7 10s.

131.—Hall-marked 9 ct. £3 10s., £4, £4 10s.   15 ct. £5 17s., £6 13s., £7 8s.

132.—Hall-marked 9 ct. £4 10s., £5, £6.   15 ct. £5 17s., £8 6s., £10.

133.—Hall-marked 9 ct. £5 15s., £4, £4 10s.   15 ct. £5 18s., £6 13s., £7 10s.

134—Hall-marked 9 ct. £6, £6 10s., £7.   15 ct. £10, £10 17s., £11 13s.

Park House, Blaengarw, South Wales.

H. MORGAN, Watch and Clock Maker, Jeweller, Etc.

Gent.'s Gold Pendants, &c.

The following illustrations do not comprise the whole the assortment I keep in stock.

Every description of Seals, Compasses, Charms, and Medals, also Gents Studs, Links, and Scarf Pins. a large Stock always ready for delivery.

Any of these patterns can be supplied in Hollow Bracket from 10/- up to 14/- and only the band stamped 9 ct.

135.—Hall-marked 9 ct. 15/-
15 ct. 48/-

136—9 ct. 30/-

137—Hall-marked 9 ct. 30/-
15 ct. 65/-

138.—Hall-marked 9 ct 24/-
15 ct. 48,-

139.—Hall-marked 9 ct. 15/-
15 ct. 33/-

140—Hall-marked 9 ct. 26/-
15 ct. 53/-

Park House, Blaengarw, South Wales.

H. MORGAN, Watch and Clock Maker, Jeweller, Etc.

Gent.'s Gold Pendants, &c.,

SUITABLE FOR STRONG AND HEAVY ALBERTS.

A smaller size than those illustrated, I can supply much cheaper.

141—Hall-marked 9 ct. 30/-
15 ct. 65/-

142—9 ct. 35/-

143—Hall-marked 9 ct. 30/-
15 ct. 59/-

144.—Hall-marked 9 ct. 30/-
15 ct. 65/-

145.—9 ct. 30/-

SUITABLE FOR LADIES.—I have a large assortment of Gold Pendants and Charms both in large and small sizes, on view, and on application, will forward samples of the same on approbation.

Park House, Blaengarw, South Wales.

H. MORGAN, Watch and Clock Maker, Jeweller, Etc.

Ladies' Gold Brooches.

NEW DESIGNED GOODS.

Velvet and Colored Satin-lined Cases for Brooches from 1/- Also Cases for Brooch and Earrings.

Any customer desiring a Set, by quoting number and design of Brooch on their order, I can easily supply the set to match complete.

Any article, however small will be forwarded to any part of the United Kingdom, safely packed and registered.

I stock an endless variety of Brooches which it would be impossible to show in this List, so I have put in what I think the most becoming designs between the large and extremely small Brooches, which will never be out of fashion. So that any order that you may have for patterns not in this list, forward on to me and I shall be pleased to execute the same.

146.—Spray Brooch, Oak and Ivy, 9 ct. 12/-
15 ct. 28/-

147.—9 ct. 13/6.
15 ct. 30/- Dead Gold.

148.—9 ct. 7/9.
15 ct. 12/6, Diamond 60/-

149 —9 ct. 14/6.
15 ct. 31/- Real Pearls.

150.—9 ct. 10/- Paste,
15 ct. 16/- Pearls.

151.—9 ct. 9/-
15 ct. 15/-

152.—9 ct. Swallow Brooch, 9/6,
15 ct. 17/6.

153.—9 ct. 14/-
15 ct. 25/-

Park House, Blaengarw, South Wales.

H. MORGAN, Watch and Clock Maker, Jeweller, Etc.

Ladies' Gold Brooches.

154.—9 ct. 21/-
15 ct. 36/- Pearls.

155.—Good Luck, 9 ct. 12/-
13 ct. 25/-

156.—9 ct. 5/-

157.—9 ct. 20/-
15 ct. 35/-

158.—9 ct. 16/-
15 ct. 30/-

159.—9 ct. 16/6
15 ct. 26/-

160.—9 ct 12/6.
15 ct. 21/-

161.—9 ct.
15 ct.

162.—9 ct. 13/-
15 ct. 26/-

163.—9 ct. 15/-
15 ct 28/-

164.—9 ct. 9/-
15 ct. 18/-

165.—9 ct. 15/-
15 ct. 29/-

166.—9 ct. 10/-
15 ct. 20/-

Park House, Blaengarw, South Wales.

H. MORGAN, Watch and Clock Maker, Jeweller, Etc.

Electro-Silver Plate.

Sterling Silver-Plate, on best white Nickel Silver, Hard Soldered.

5468—Egg Frame,
E.P.N S. 38/-

G7841—Biscuit Boat
E.P. quality B 28/-   ai 35/-

G5072—Grape Stand gilt inside
E.P.N.S. 40/-

6879½—With China or Glass Dish
E.P.N.S. 33/-

6289—Fruit Basket, Quality B.
E.P.N.S. 35/-

7593—Jam Dish,
E.P. 35/-

5265—Toast Rack, mounted on oak,
28/-   E.P.N.S. 18/-

5481—Jam Dish,
E.P.N.S. 11/6

Park House, Blaengarw, South Wales.

H. MORGAN, Watch and Clock Maker, Jeweller, Etc.

Electro-Silver Plate,
TEA AND COFFEE SERVICES, ONLY 87/- PER SET.

7480—Coffee Pot, P. 28/-

G7843—Biscuit Boat E.P. 20/-

7840—Teapot, 5 gills, Quality B 24/6.

7857—Breakfast Cruet, E.P.N.S. 20/-

7801—Breakfast Cruet. E.P. 16/6

7872—Breakfast Cruet, E.P. 12/6

7840—Sugar Basin, 18/6.

7558—Jam Dish, E.P. 13/-

7480—Cream Jug, 16/6.

Park House, Blaengarw, South Wales.

H. MORGAN, Watch and Clock Maker, Jeweller, Etc.

Electro-Silver Plate.

Sterling Silver Plate on best White Nickel Silver.

9184½—Coffee Pot, 5 gills, 75/-   9184½—Teapot, 4 gills, 64/-   9184½—Cream Jug, 15/-
9184½—Sugar Basin, 25/-
The Massive Set, Complete, only £8 19s.

9887—Cream and Sugar Stand, China 30/6

G7815—Biscuit, Butter and Cheese Basket, 26/6

9633—Teapot, 17/6   7798—Breakfast Cruet, E.P. 14/-   9622—Teapot, 24/-

Park House, Blaengarw, South Wales.

H. MORGAN, Watch and Clock Maker, Jeweller, Etc.

Dinner, Luncheon, and Breakfast Cruets,
Best Silver Plate on Barmaroid Metal or Nevada Silver, and E.P.N.S.

8761—Dinner Cruet, 4 bottles,
E.P. 30/-

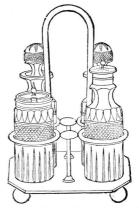

8743—Dinner Cruet, 4 bottles,
E.P. 35/-

7871—Luncheon Cruet,
E.P.N.S. 14/6.

7873—Breakfast Cruet,
E.P. 10/6, quality D.

2334—E.P. Nut Crackers, Ebony Handles, 4/-
With Velvet-lined Case, 6/-

8682—E.P. Dinner Cruet,
4 bottles 25/- quality D.

Park House, Blaengarw, South Wales.

H. MORGAN, Watch and Clock Maker, Jeweller, Etc.

Best E.S. Plated Knives, Spoons and Forks,
Same Patterns can be supplied in Sterling Silver, Hall-marked.

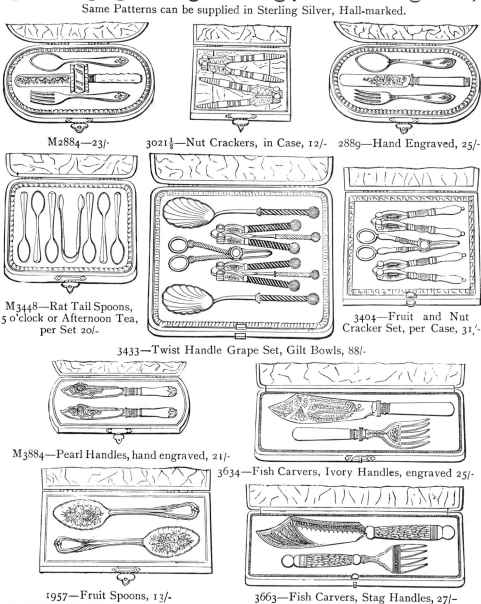

M2884—23/-

3021½—Nut Crackers, in Case, 12/-

2889—Hand Engraved, 25/-

M3448—Rat Tail Spoons, 5 o'clock or Afternoon Tea, per Set 20/-

3433—Twist Handle Grape Set, Gilt Bowls, 88/-

3404—Fruit and Nut Cracker Set, per Case, 31/-

M3884—Pearl Handles, hand engraved, 21/-

3634—Fish Carvers, Ivory Handles, engraved 25/-

1957—Fruit Spoons, 13/-

3663—Fish Carvers, Stag Handles, 27/-

Park House, Blaengarw, South Wales.

H. MORGAN, Watch and Clock Maker, Jeweller, Etc.

Best Electro-Plated Nickel Silver.

Every article stamped E.P.N.S.   There are many firms supplying very thin Electro at same prices as I sell them, but are not stamped.

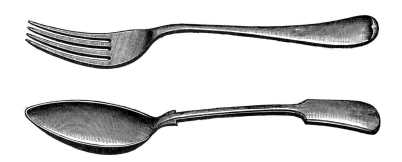

	A	AI	B	C
Table Spoons	52/-	39/-	26/-	22/-
Dessert Spoons	34/6	26/-	18/6	16/-
Table Forks	52/-	39/-	26/-	22/-
Dessert Forks	34/6	26/-	18/6	16/-
Tea Spoons	21/9	16/6	11/9	8/-
Salt Spoons	21/6	15/6	9/6	8/-
Sugar Tongs	44/-	52/-	24/-	18/-
Mustard Spoons	10/-	8/6	6/6	5/-

Larger Size 3/- extra per dozen.
Smaller Size 2/6 less per dozen.

NEW REGISTERED MERIDIAN PATTERN IN NEVADA SILVER,
Warranted to wear silvery white through.

	Per dozen.
Table Spoons	13/-
Dessert Spoons	10/-
Table Forks	13/-
Dessert Forks	10/-
Tea Spoons	5/-
Sugar Fongs	24/6
Salt Spoons	5/-
Mustard Spoons	5/-

SUPERIOR KNIVES ONLY.
Per dozen.
Table Knives, Ivory handles 50/- 55/- 60/-
Dessert   ,,   ,,   40/- 45/- 50/-
Table   ,, white handles 10/- 13/- 17/- 21/-
Dessert   ,,   ,,   7/6 8/6 10/6 16/-
Table   ,, Imitation Ivory 16/- 18/- 21/-
Dessert   ,,   ,,   ,; 12/- 14/- 16/-
Table   ,, Stain horn handles 15/- 18/- 21/-
Dessert   ,,   ,,   ,,   12/- 14/- 17/-

KNIVES AND FORKS,
In Sets, per dozen:
Black handles   8/- 10/- 12/- 14/- 16/- 18/-
Buck Horn handles   ...   12/- 14/- 16/-
White Bone Handles   ... 8/- 10/- 12/- 14/-
Polished Horn with Tips ... 12/- 14/- 16/-

CARVERS, per Set:
Patent Ivory Handles, quality, 10/- 12/- 14/-
Buck Horn or White 4/6 5/6 6/6 7/6 8/6

Park House, Blaengarw, South Wales.

H. MORGAN, Watch and Clock Maker, Jeweller, Etc.

Marble Clocks, Time-pieces and Alarums.

These are beautifully Carved Cabinet with Good Sound Movement, also 30-Hour Nickel Alarums, Extra Make.

167—" Joker," Striker, Gold Gilt Nickel-Plated Front 18/-

168—" Mikado," Gold Gilt Front and Handle, 1-day Alarum, 16/-

169—Imitation Marble, (44) 8-day, strikes on gong hours and half-hours.

170—8-day Time-piece, Polished Wood Case, 24/-

Park House, Blaengarw, South Wales.

H. MORGAN, Watch and Clock Maker, Jeweller, Etc.

Marble Clocks, Time-pieces and Alarums.

171—"Bangor," Walnut Case, Strikes on Cathedral Gong ½-hours, 8-day 30/-

172—"Concord," Walnut Case, Strikes on Cathedral Gong ½-hours, 8-day, 44/-

173—Walnut Case, good timekeeper, good make, 8/6

174—Imitation Marble Case Time-piece, 13/-

Park House, Blaengarw, South Wales.

H. MORGAN, Watch and Clock Maker, Jeweller, Etc.

Marble Clocks, Time-pieces and Alarums.

175—"Echo" Alarum, Nickel Plated, 10/-

176—"Elk" Alarum, Nickel Plated 13/-

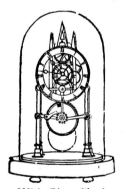

177—With Glass Shade, 25/-

178—Nickel Alarum, 4/6

179—Nickel Alarum Clock, 16/-

Park House, Blaengarw, South Wales.

H. MORGAN, Watch and Clock Maker, Jeweller, Etc.

Marble Clocks, Time-pieces and Alarums.

180—Sharp "Gothic," Rosewood or Walnut Veneered Polished Case, Striker, 22/-

181—"Gordon," Walnut Case, 8-day, Strikes ½ hours on Cathedral Gong, 27/-

182—14-day White Marble Clock, with Shade, 40/-

183—Wood Case Timepiece, 8/6

Park House, Blaengarw, South Wales.

H. MORGAN, Watch and Clock Maker, Jeweller, Etc.

Marble Clocks, Time-pieces and Alarums.

184—"Albert," Rosewood or Walnut Case, Strikes 25/-

185—"Selma," Old Oak, 8 day, strikes hours and ½-hours, 45/-

186—(4) Office Clock, Old Oak, 8-day, 34/-

187—(5) Office Clock, 8-day, 41/-

Park House, Blaengarw, South Wales.

H. MORGAN, Watch and Clock Maker, Jeweller, Etc.

Marble Clocks, Time-pieces and Alarums.

189—8-day Striking Clock, with line and dot, £1 10s.

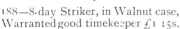

188—8-day Striker, in Walnut case, Warranted good timekeeper £1 15s.

190—Large Clock, solid heavy walnut case, best movement 10/-   191—Lever Clock, Nickel Case, 12/-   192—4-day Striking Clock, 20/- beautiful Wood Case.

Park House, Blaengarw, South Wales.

H. MORGAN, Watch and Clock Maker, Jeweller, Etc.

Bracelets, Necklets, Presentation Tea Pots, Tea and Coffee Sets.

193—15 ct. Bangle, £6 15s.

194—15 ct. Pearl Bangle, £3 10s.

195—9 ct. Paste Bangle, 30/-

196—9 ct. Bracelet, common Pearl, £10.

197—Tea Pot, 36/-

198—Very Massive Tea Pot, 95/-

199—Presentation Set Complete, £20.

200—Tea Set with Tray, £15.

Park House, Blaengarw, South Wales.

H. MORGAN, Watch and Clock Maker, Jeweller, Etc.

Tureens, Butter Coolers, Cheese Stands, &c.

201—Chop Dish, £6 10s.

204—Soup Dish, £8 8s.

202—Dessert Stand, £4 10s.

205—Biscuit Pot, China, 14/-

203—Chop Dish, £7.

206—Epergne for Flowers, £6 5s.

Park House, Blaengarw, South Wales.

H. MORGAN, Watch and Clock Maker, Jeweller, Etc.

Challenge and Presentation Cups.

BEST STERLING SILVER-PLATED, HAND ENGRAVED, Inferior quality half-price.

207—£4 10s. 207A—£7 15s. 207B—£12. 207C—£6. 207D—£3 15s.

208—£2 15s. 209—£20. 210—£2.

211—£4. 212— 213—£3 15s.

Park House, Blaengarw, South Wales.

H. MORGAN, Watch and Clock Maker, Jeweller, Etc.

Fruit Stand, Barometers, Opera and Field Glasses, Spectacles, Folders, &c.

14—14-day Timepiece, and Aneroid Barometer, £3 5s.

215—

216—Wheel Barometer, 20/- Aneroid Barometer, 40/-

17—Opera Glasses from 8/-, 10/-, 13/-, 18/-, and 21/-

218—Best Pebbles from 5/6 to 10/- any Lenses

19—Microscope, /-, 25/-, and 30/-

220—Steel Folders, 2/6, 8/6, 4/6, to 10/-

221—Telescope with Stand, for long distances 4 drawers, £3 10s.

Park House, Blaengarw, South Wales.

H. MORGAN, Watch and Clock Maker, Jeweller, Etc.

Musical Instruments, and Bicycles.

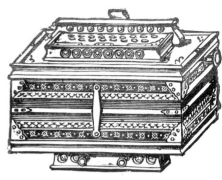

223—Iron Frame Piano, well finished from £20 to £50, full size.

222—Solid Walnut Organ, 9 Stops, 2 or 3 Octave Couplers, 1 Octave Sub-Bass, height 5 ft., length, 3-ft. 9-in., depth 1-ft. 8-in., £20.

224—Musical box £2 to £12.

225—Double Bellows Accordian, cannot be beaten at the price, 11/-
The Chromatic Accordian, very highly finished, 4 Bass notes, 19 keys (bone), very powerful tone, £1 6s.

226—American Organ and Harmonium, from £15 to £40. Cottage Harmonium, from £6, very powerful.

227—Concertina English make, 20 bone keys, £2
Latest Improved Concertina, 30 keys, tempered Steel Reeds, £4 15s.

228—Orguinette, with 8 tunes, complete Large size, 60/-

229—Organ Accordian, 3 Octaves, in wood box, suitable for sacred Songs, £4.

230—Cornet 40/- to 60/-

231—Best Weldless Steel Tube Bicycle, ball bearings to every part, Cushion tyres £9, Pneumatic Tyres £12.

Park House, Blaengarw, South Wales.